SOLUTIONS MANUAL

for the

ELECTRICAL ENGINEERING REVIEW MANUAL

Fourth Edition

Including Solutions to the Sample Exam

Raymond B. Yarbrough

In the *ENGINEERING REVIEW MANUAL SERIES:*

Engineer-In-Training Review Manual
Quick Reference Cards for the E-I-T Exam
Mini-Exams for the E-I-T Exam
Civil Engineering Review Manual
Seismic Design for the California Civil P.E. Exam
Timber Design for the Civil P.E. Exam
Mechanical Engineering Review Manual
Electrical Engineering Review Manual
Chemical Engineering Review Manual
Expanded Interest Tables
Engineering Law, Design Liability, and Professional Ethics

Distributed by: Professional Publications, Inc.
Post Office Box 199
Department 77
San Carlos, CA 94070
(415) 593-9119

SOLUTIONS MANUAL for the ELECTRICAL ENGINEERING REVIEW MANUAL

Printed in the United States of America

ISBN 0-932276-41-5

Professional Engineering Registration Program
Post Office Box 911, San Carlos, CA 94070

Current printing of this edition (last number) 5 4 3

TABLE OF CONTENTS

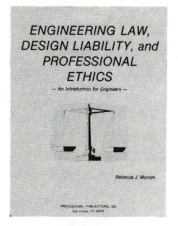

1. 1101.011

$1\times2^3=8$, $1\times2^2=4$, $0\times2^1=0$, $1\times2^0=1$; total $= 13$

$0\times2^{-1}=0$, $1\times2^{-2}=.25$, $1\times2^{-3}=.125$; total $= .375$

$$(1101.011)_2=(13.375)_{10}$$

2. $(4+j5)\times(2+j3)=8+j12+j10+j^2 15$
$=8-15+j(12+10)=-7+j22$

3. $(4+j5)\times(4-j5)=16+j20-j20-j^2 25$
$=16+25=41$

$\left(\sqrt{4^2+5^2}\angle\tan^{-1}\frac{5}{4}\right)\times\left(\sqrt{4^2+5^2}\angle\tan^{-1}\frac{-5}{4}\right)$
$=(4^2+5^2)\left[\angle\tan^{-1}\frac{5}{4}-\angle\tan^{-1}\frac{5}{4}\right]=41$

4. $\dfrac{1}{2-j3}\times\dfrac{2+j3}{2+j3}=\dfrac{2+j3}{4+9}=.154+j.231$

or $\dfrac{1}{\sqrt{2^2+3^2}\angle-\tan^{-1}\frac{3}{2}}=\dfrac{1}{\sqrt{13}}\angle\tan^{-1}\frac{3}{2}=.277\angle56.3^\circ$

5. $\dfrac{4+j5}{2-j3}\times\dfrac{2+j3}{2+j3}=\dfrac{-7+j22}{13}$
$=-.538+j1.692$

or $\dfrac{\sqrt{4^2+5^2}\angle\tan^{-1}5/4}{\sqrt{2^2+3^2}\angle\tan^{-1}-3/2}=\sqrt{\dfrac{41}{13}}\left[\angle\tan^{-1}\frac{5}{4}+\angle\tan^{-1}\frac{3}{2}\right]$
$=1.78\angle107.7^\circ$

6. $\dfrac{6.40e^{j.896}}{3.61e^{-j.983}}=1.77e^{j(.896+.983)}$
$=1.77e^{j(1.879)}$

$1.879\text{ rad}\times\dfrac{180^\circ}{\pi\text{ rad}}=107.7^\circ$

the difference due to a rounding error.

7. $\dfrac{1}{x(x+1)}=\dfrac{A}{x}+\dfrac{B}{x+1}=\dfrac{A(x+1)+Bx}{x(x+1)}$

so $(A+B)x+A=1$ $\quad A=1,\ B=-A$

8. $\dfrac{4x^2+5x+2}{x^2(x+1)}=\dfrac{A}{x}+\dfrac{B}{x^2}+\dfrac{C}{x+1}$
$=\dfrac{Ax(x+1)+B(x+1)+Cx^2}{x^2(x+1)}$
$=\dfrac{(A+C)x^2+(A+B)x+B}{x^2(x+1)}$

then
$A+C=4$; $A+B=5$; $B=2$
solving
$A=3,\ C=1$

9. $\dfrac{5x^2+11x+20}{x(x^2+2x+5)}=\dfrac{A}{x}+\dfrac{Bx+C}{x^2+2x+5}$

$(A+B)x^2=5x^2$
$(2A+C)x=11$
$5A=20$

$A=4$
$B=1$
$C=3$

10.
$$\begin{array}{r} s+2 \\ s^2+2s+1\,\overline{)\,s^3+4s^2+6s+2} \\ \underline{s^3+2s^2+s} \\ 2s^2+5s+2 \\ \underline{2s^2+4s+2} \\ s \end{array}$$

$\dfrac{s^3+4s^2+6s+2}{s^2+2s+1}=s+2+\dfrac{s}{s^2+2s+1}$

11. let $x=\frac{\theta}{2}$ so $\theta=2x$
$\cos 2x=2\cos^2 x-1=1-2\sin^2 x$
$\cos^2\frac{\theta}{2}=\frac{1}{2}(1+\cos\theta)$
$\sin^2\frac{\theta}{2}=\frac{1}{2}(1-\cos\theta)$

12. (a) by inspection $V=50$

(b) left segment and right segment intersect at $(I,V)=(5,90)$

left: $\frac{V-90}{I-5} = \frac{100-90}{0-5} = -2$

$V = -2I + 100$

right: $\frac{V-90}{I-5} = \frac{0-90}{10-5} = -18$

$V = -18I + 180$

(c)

left: $I=0$

right straight segment would intersect axis at $I=2.4$:

$$\frac{V-0}{I-2.4} = \frac{100-0}{4.5-2.4} = 47.6$$

$V = 47.6I - 114.2$

13. $\frac{du}{dx} = 2\cos x - 3\sin x = 0$

$\therefore \tan x = \frac{2}{3}$; $x = 33.7°, -146°$

$-146°$ is not between limits, then solution is $x = 33.7°$

$\frac{d^2u}{dx^2} = -2\sin x - 3\cos x = -2.77$

therefore u is maximum at $x = 33.7°$

14. $\bar{u} = \frac{\int_0^1 e^{-2x}\cos \pi x \, dx}{1-0}$ [see eq 1.64]

$$= \frac{e^{-2x}}{2^2+\pi^2}\left(-2\cos \pi x + \pi \sin \pi x\right)\Big|_{x=0}^{x=1}$$

$$= \frac{e^{-2}(-2\cos\pi + \pi \overset{0}{\sin\pi}) - e^{0}(-2\cos 0 + \pi \overset{0}{\sin 0})}{4+\pi^2}$$

$$= \frac{2e^{-2}+2}{4+\pi^2} = .164$$

15. $b=.01$; $C=.0025$, $d=0$, $x_0 = v_0 = 100$

$\frac{d}{c} = 0$

$v(t) = 0 + (100-0)e^{-\frac{.0025}{.01}t}$

$= 100e^{-.25t}$

16. $i = .5 + Ae^{-4t} + Be^{-t}$

$\frac{di}{dt} = -4Ae^{-4t} - Be^{-t}$ } see ex. 1.11

at $t=0$: $i = 1.7 = .5 + A + B$; $\frac{di}{dt} = 0 = -4A - B$ } $A=-.4$, $B=1.6$

$i(t) = .5 - .4e^{-4t} + 1.6e^{-t}$

17. $i = .25 + Ae^{-t} + B\,e^{-t}$

$\frac{di}{dt} = -Ae^{-t} + Be^{-t} - Bte^{-t}$ } see ex 1.12

at $t=0$: $i(0) = .8 = .25 + A \;\therefore\; A = .55$

$\frac{di(0)}{dt} = 0 = -A + B \;\therefore\; B = .55$

$i(t) = .25 + .55(1+t)e^{-t}$

18. $b^2 - 4ac = -16$ $\alpha = \frac{3}{2} = 1.5$

$\beta = \frac{\sqrt{16}}{2} = 2$

$v(t) = -\frac{12.5}{6.25} + Ae^{-1.5t}\cos 2t + Be^{-1.5t}\sin 2t$

$\frac{dv}{dt} = -1.5Ae^{-1.5t}\cos 2t - 1.5Be^{-1.5t}\sin 2t - 2Ae^{-1.5t}\sin 2t + 2Be^{-1.5t}\cos 2t$

at $t=0$: $v(0) = 10 = -2 + A \;\therefore\; A = 12$

$\frac{dv(0)}{dt} = -1.5A + 2B = 0 \;\therefore\; B = 9$

$v(t) = -2 + 12e^{-1.5t}\cos 2t + 9e^{-1.5t}\sin 2t$

19.

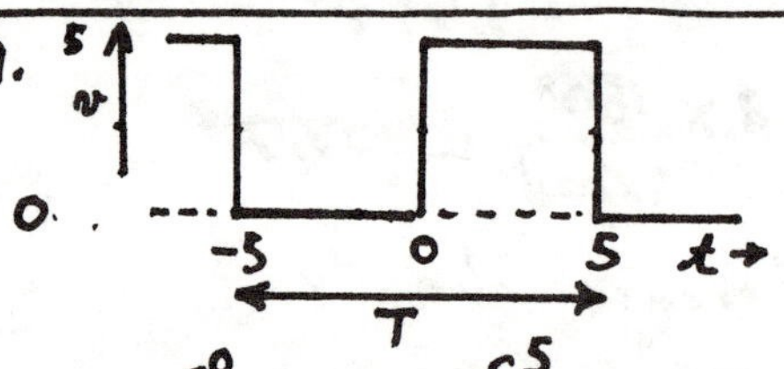

$$a_0 = \frac{2}{10}\int_{-5}^{0} 0\,dt + \frac{2}{10}\int_{0}^{5} 5\,dt = 5$$

19. continued

$$a_n = \frac{2}{10}\int_0^5 5\cos\frac{2\pi}{T}nt\,dt \qquad T=10$$

$$= \frac{1}{5}\,\frac{10}{2\pi n}5\sin\frac{2\pi}{10}nt\Big|_0^5 \equiv 0$$

$$b_n = \frac{2}{10}\int_0^5 5\sin\frac{2\pi n}{10}t\,dt$$

$$= \frac{-10}{2\pi n}\cos\frac{2\pi n}{10}t\Big|_0^5 = \frac{5}{\pi n}(1-\cos\pi n)$$

$$\cos\pi n = (-1)^n$$

$$b_1 = \frac{10}{\pi},\ b_2 = 0,\ b_3 = \frac{10}{3\pi},\ b_4 = 0 \text{ etc}$$

$$v(t) = \frac{5}{2} + \sum_{n=1}^{\infty}\frac{5}{\pi n}(1-\cos\pi n)\sin\frac{2\pi n}{10}t$$

to eliminate the zero terms

$$b_n = \frac{10}{(2m-1)\pi} \quad \text{for } n = 1, 3, 5 \text{ when } m = 1, 2, 3 \text{ etc}$$

then

$$v(t) = 2.5 + \sum_{m=1}^{\infty}\frac{10}{(2m-1)\pi}\sin\frac{(2m-1)\pi}{5}t$$

20. Direct term-by-term transform:

$$\frac{.02}{s} = 10^{-6}\left[s^2I - \frac{di(0)}{dt} - s\,i(0)\right] + 2\times10^{-4}\left[sI - i(0)\right] + I$$

$$\frac{.02}{s} = 10^{-6}s^2I + 2\times10^{-4}sI + I - \left[10^{-6}s(.5) + 10^{-6}(50) + 2\times10^{-4}(.5)\right]$$

$$\left[s^2 + 200s + 10^6\right]I = \frac{2\times10^4}{s} + 150 + .5s = \frac{.5s^2 + 150s + 2\times10^4}{s}$$

21.

$$I(s) = .5\,\frac{s^2 + 300s + 4\times10^4}{s(s^2 + 200s + 10^6)}$$

22.

$$I(s) = \frac{A}{s} + \frac{Bs + C}{s^2 + 200s + 10^6} = \frac{(A+B)s^2 + (200A+C)s^2 + 10^6A}{s(s^2 + 200s + 10^6)}$$

matching

$$A+B = .5\ ;\ 200A + C = 150\ ;\ 10^6A = 2\times10^4$$

$$A = .02,\ B = .48,\ C = 146$$

$$I(s) = \frac{.02}{s} + \frac{.48s + 146}{s^2 + 200s + 10^6}$$

23. $b^2 - 4ac = -3.96\times10^6$

so system is underdamped

the table uses the forms

$$\frac{s+a}{(s+a)^2 + b^2} \quad \text{and} \quad \frac{b}{(s+a)^2 + b^2}$$

so

$$\frac{.48s + 146}{s^2 + 200s + 10^6} = .48\,\frac{s + a + bk}{(s+100)^2 + 995^2}$$

$$a = 100,\ b = 995$$

$$.48(100 + bk) = 146\ :\ bk = 204$$

$$K = .205$$

then

$$\frac{.48s + 146}{s^2 + 200s + 10^6} = .48\,\frac{s + 100}{(s+100)^2 + 995^2} + .48\times.205\,\frac{995}{(s+100)^2 + 995^2}$$

then

$$i(t) = .02 + .48e^{-100t}\cos 995t + .098e^{-100t}\sin 995t$$

1. $\rho_{20} = 1.08 \times 10^{-6}\ \Omega m$; $\alpha_{20} = .017/°C$

$$A = \frac{\pi}{4}\left[\frac{\sqrt{100}}{1000}\text{ in.} \times .0254\frac{m}{in}\right]^2$$

$$= 5.08 \times 10^{-8}\ m^2$$

$$R_{100} = \frac{\rho_{20}[1 + .017(100-20)]\ell}{A} = 1\ \Omega$$

so $\ell = .0199\ m$

$$R_{20} = \frac{\rho_{20}\ell}{A} = 0.424\ \Omega$$

2. $v_c(0^+) = v_c(0^-) = \dfrac{V_s R_2}{R_1 + R_2}$

$$i_c(0^+) = -\frac{V_c(0^+)}{R_2} = \frac{-V_s}{R_1 + R_2} \quad (a)$$

$$\frac{dv_c(0^+)}{dt} = \frac{i_c(0^+)}{C} = \frac{-V_s}{C(R_1 + R_2)} \quad (b)$$

$$v_c(\infty) = 0 \quad (c)$$

3. $i_L(0^+) = i_L(0^-) = V_s/R_1 \quad (a)$

$$\frac{di_L(0^+)}{dt} = -\frac{V_s}{R_1}\frac{R_2}{L} \quad (b)$$

$$v_L(0^+) = -\frac{V_s}{R_1}R_2 \quad (c)$$

4. Let 125T winding be #1: $N_1 = 125$
500T winding be #2: $N_2 = 500$
$N_3 = 36$

$$\frac{V_1}{N_1} = \frac{120}{125} = \frac{V_2}{500} = \frac{V_3}{36} \quad \therefore V_2 = 480V,\ V_3 = 34.56V$$

$$i_2 = -\frac{480}{60} = -8A \ ;\ i_3 = -\frac{34.56}{3} = -11.52\ A$$

$$\Sigma NI = 125 I_1 + 500(-8) + 36(-11.52)$$

$$I_1 = 35.32A$$

ratings must exceed:

35.32A for 125T
11.52A for 36T
8.0 A for 500T

5. slope $= \dfrac{10-9}{0-6} = -\dfrac{1}{6}$

$$V = -\tfrac{1}{6}I + 10 \quad \text{or} \quad I = -6V + 60$$

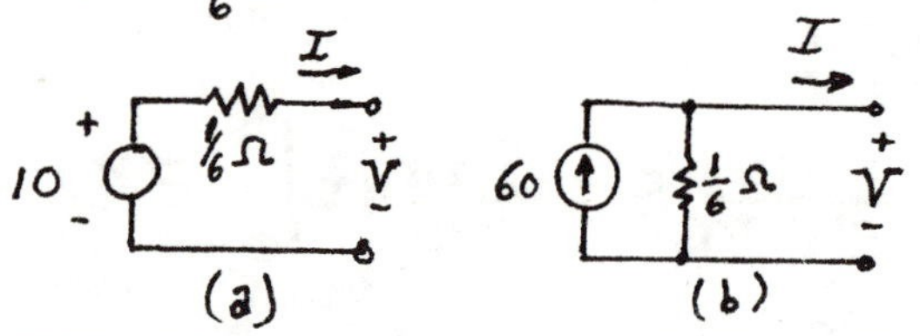

6. $\%Reg = \dfrac{V_{NL} - V_{FL}}{V_{FL}} \times 100 = 1\%$

$$V_{NL} = 1.01 V_{FL} = 5.05V = V_{Th}$$

$$\frac{V_{NL} - V_{FL}}{I_{FL}} = \frac{.05}{25} = .002\ \Omega = R_{Th}$$

$$I_N = \frac{5.05}{.002} = 2525A$$

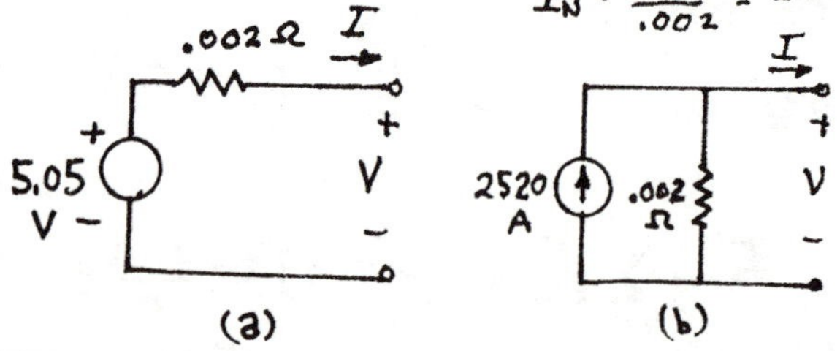

7.

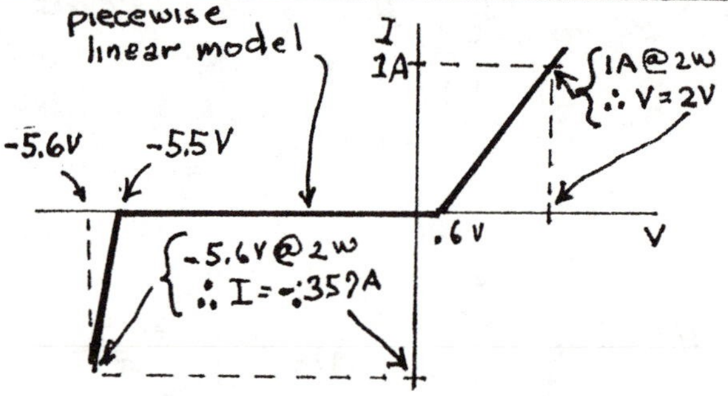

(a) Reverse breakdown

$$\frac{V-(-5.5)}{I-0} = \frac{-5.6-(-5.5)}{-.357-0}$$

$$V = -5.5 + .28I$$

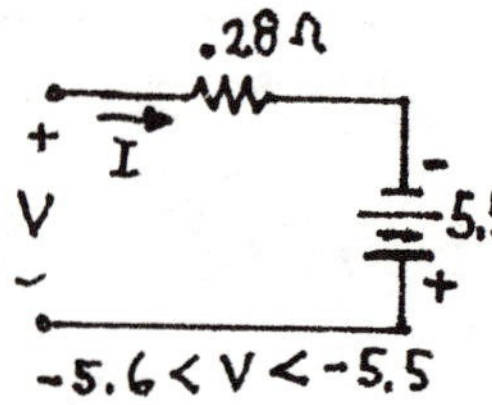

(b) $-5.5 < V < .6$

$$I = 0$$

(open circuit)

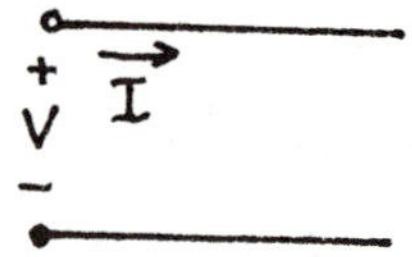

(c) forward bias $V > .6$

$$\frac{V - .6}{I - 0} = \frac{2 - .6}{1 - 0}$$

$$V = .6 + 1.4I$$

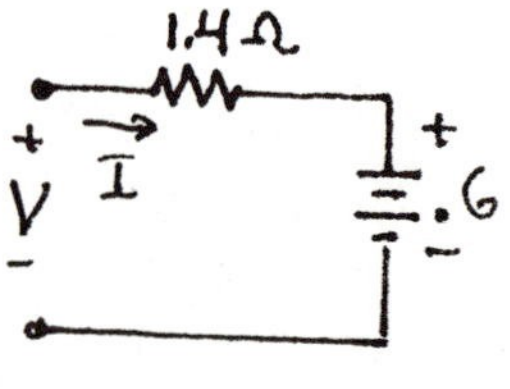

8.

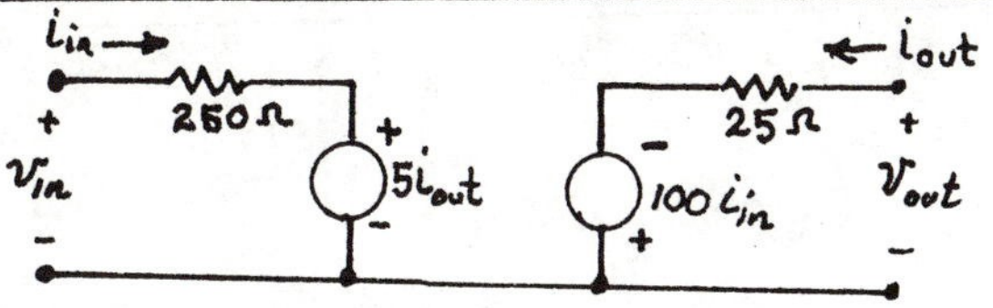

9. $v_1 = v_{eb}$, $i_1 = i_e$, $v_2 = v_{cb}$, $i_2 = i_c$

current through $r_b \downarrow$ is $i_e + i_c$

current through $r_c \leftarrow$ is $i_c + \alpha i_e$

$$v_{eb} = i_e r_e + (i_e + i_c) r_b$$

$$v_{cb} = (i_c + \alpha i_e) r_c + (i_e + i_c) r_b$$

or

$$v_{eb} = (r_e + r_b) i_e + r_b i_c$$

$$v_{cb} = (r_b + \alpha r_c) i_e + (r_b + r_c) i_c$$

$$\Delta = \begin{vmatrix} (r_e + r_b) & r_b \\ (r_b + \alpha r_c) & (r_b + r_c) \end{vmatrix} = r_e(r_b + r_c) + r_b r_c (1-\alpha)$$

$$i_e = \frac{\begin{vmatrix} v_{eb} & r_b \\ v_{cb} & (r_b + r_c) \end{vmatrix}}{\Delta} = \frac{(r_b + r_c) v_{eb} - r_b v_{cb}}{\Delta}$$

$$i_c = \frac{\begin{vmatrix} (r_e + r_b) & v_{eb} \\ (r_b + \alpha r_c) & v_{cb} \end{vmatrix}}{\Delta} = \frac{(r_e + r_b) v_{cb} - (r_b + \alpha r_c) v_{eb}}{\Delta}$$

then

$$y_{11} = y_{ib} = \frac{r_b + r_c}{r_e(r_b + r_c) + r_b r_c (1-\alpha)}$$

$$y_{12} = y_{rb} = \frac{-r_b}{r_e(r_b + r_c) + r_b r_c (1-\alpha)}$$

$$y_{21} = y_{fb} = \frac{-(r_b + \alpha r_c)}{r_e(r_b + r_c) + r_b r_c (1-\alpha)}$$

$$y_{22} = y_{ob} = \frac{r_e + r_b}{r_e(r_e + r_b) + r_b r_c (1-\alpha)}$$

10. from ex 2.11

$$i = .7 + .0192 I_L \ \mu A$$

then $V_L = 10^4 i = .007 + (1.92 \times 10^{-4}) I_L$

LOAD LINE : $5 = V_f + 10^4 i$

which is valid for

$$0 < I_L < 2500 \text{ f.c.}$$

11. $\dfrac{V_0 - .95}{T - 0} = \dfrac{.6 - .95}{100 - 0} = -.0035$

$V_0 = .95 - .0035T$; $v_f = V_0 + .3 i_f$

then $v_f = .95 - .0035T + .3 i_f$

12. $V_y = \dfrac{I_x B_z}{w n q} = \dfrac{.1 B_z}{(5 \times 10^{-3}) \times 10^{19} \times (1.6 \times 10^{-19})} = 12.5 B_z$

$V_y = 12.5 \mu_0 H_z = 12.5 \times 4\pi \times 10^{-7} H_z$

$= 1.57 \times 10^{-5} H_z$

13. $i_b = -\dfrac{v_{in}}{1500}$; $i_{s.c.} = \dfrac{N_1}{N_2}(-50 i_b) = \dfrac{N_1}{N_2} \dfrac{v_{in}}{30}$

$v_{o.c.} = \dfrac{N_2}{N_1}(-50 i_b \times 10^4) = \dfrac{N_2}{N_1} \dfrac{10^3}{3} v_{in}$

$Z_{NORTON} = \dfrac{v_{o.c.}}{i_{s.c.}} = \left(\dfrac{N_2}{N_1}\right)^2 \times 10^4$

$i_{NORTON} = i_{s.c.} = \dfrac{N_1}{N_2} \dfrac{v_{in}}{30}$

14. $R_L = Z_{th} = \left(\dfrac{N_2}{N_1}\right)^2 \times 10^4 = 8 \ \Omega$

$\dfrac{N_1}{N_2} = \sqrt{\dfrac{10^4}{8}} = 35.35$

15. $Z_{th} = \dfrac{jX_2(50 + jX_1)}{50 + j(X_1 + X_2)}$; $v_{th} = \dfrac{jX_2 v_s}{50 + j(X_1 + X_2)}$

for d.c. $-X_2 \to \infty$, $X_1 \to 0$

$Z_{th}(\text{d.c.}) = \dfrac{50 X_2}{X_2} = 50$, $v_{th}(dc) = v_s(dc)$

for 2236 r/s component

$X_2 = -447.2$, $X_1 = 447.2$

$Z_{th}(2236) = \dfrac{-j447.2(50 + j447.2)}{50 + j(-447.2 + 447.2)}$

$= 4000 - j447.2$

$v_{th}(2236) = \dfrac{-j447.2}{50} v_s(2236) = j8.94 v_s(2236)$

$v_{100} = v_{th} \dfrac{100}{100 + Z_{th}}$

$v_{100}(dc) = 20 \times \dfrac{100}{150} = 13.3$

$v_{100}(2236) = \dfrac{j8.94 \times 5 \times 100}{4100 - j447.2} = 1.084 \angle -83.8°$

$v_{100} = 13.3 + 1.084 \cos(2236t - 83.8°)$

16. $r_b + R_e = 210$, $-\beta r_c + R_e \to -1.25\times10^5$

$r_c + R_e = 2510$; $i_2 = -10^{-3} v_{out}$

(a) $v_{in} = 210\, i_1 + 10(-10^{-3} v_{out})$

(b) $v_{out} = -1.25\times10^{5}\, i_1 + 2510(-10^{-3} v_{out})$

from (b) $i_1 = -2.808\times10^{-5} v_{out}$

into (a) $v_{in} = -1.59\times10^{-2} v_{out}$

then $\dfrac{V_{out}}{V_{in}} = -62.9$

17.

$$\left.\begin{aligned} i_1 &= \frac{v_{in} - v_{out}}{R_f} \\ i_2 &= \frac{v_{out} - v_{in}}{R_f} \end{aligned}\right\} \begin{bmatrix} i_1 \\ i_2 \end{bmatrix} = \begin{bmatrix} \frac{1}{R_f} & -\frac{1}{R_f} \\ -\frac{1}{R_f} & \frac{1}{R_f} \end{bmatrix} \times \begin{bmatrix} v_{in} \\ v_{out} \end{bmatrix}$$

$$\left.\begin{aligned} i_1' &= \frac{v_{in}}{r_b} \\ i_2' &= g_m v_{in} + \frac{v_{out}}{r_c} \end{aligned}\right\} \begin{bmatrix} i_1' \\ i_2' \end{bmatrix} = \begin{bmatrix} \frac{1}{r_b} & 0 \\ g_m & \frac{1}{r_c} \end{bmatrix} \times \begin{bmatrix} v_{in} \\ v_{out} \end{bmatrix}$$

$$\begin{bmatrix} i_{in} \\ i_{out} \end{bmatrix} = \begin{bmatrix} i_1 \\ i_2 \end{bmatrix} + \begin{bmatrix} i_1' \\ i_2' \end{bmatrix} = \begin{bmatrix} \left(\frac{1}{R_f} + \frac{1}{r_b}\right) & -\frac{1}{R_f} \\ \left(g_m - \frac{1}{R_f}\right) & \left(\frac{1}{R_f} + \frac{1}{r_c}\right) \end{bmatrix} \times \begin{bmatrix} v_{in} \\ v_{out} \end{bmatrix}$$

CONCENTRATES

1. From the 100Ω load the Thevenin impedance is : $Z_{Th} = \dfrac{Z_2(50 + Z_1)}{50 + Z_1 + Z_2}$

for maximum power transfer $Z_{Th} = Z_L^* = 100$

Z_1 & Z_2 must be reactive so they do not absorb power

let $Z_1 = jX_1$ and $Z_2 = jX_2$. Making these substitutions and cross multiplying: $5000 + j100(X_1 + X_2) = j50X_2 + j^2(X_1X_2) = -X_1X_2 + j50X_2$

Equating Real parts : $-X_1X_2 = 5000$

Equating Imaginary parts : $50X_2 = 100(X_1 + X_2)$ or $X_2 = -2X_1$

then $2X_1^2 = 5000$, so $X_1 = \pm 50$ and $X_2 = \mp 100$

for $X_1 = 50 = 2000L_1$, $X_2 = -1000 = \dfrac{-1}{2000C_2}$ $\quad L_1 = .025\text{ H}$, $C_2 = 5\mu\text{F}$

for $X_1 = -50 = \dfrac{-1}{2000C_1}$, $X_2 = 1000 = 2000L_2$ $\quad C_1 = 10\mu\text{F}$, $L_2 = .05\text{ H}$

2. convert V_s in series with 50Ω to $I_s = \dfrac{V_s}{50}$ in parallel with 50Ω

(column → 1, 2, 3, 4; ↓ row 1, 2, 3, 4)

$$\begin{bmatrix} -50i_1 \\ 50i_2 \\ 50(i_1 - i_2) \\ \frac{V_s}{50} \end{bmatrix} = \begin{bmatrix} \left(\frac{1}{1000} + \frac{1}{1000}\right) & 0 & -\frac{1}{1000} & 0 \\ 0 & \left(\frac{1}{1000} + \frac{1}{1000}\right) & -\frac{1}{1000} & 0 \\ -\frac{1}{1000} & -\frac{1}{1000} & \left(\frac{1}{1000} + \frac{1}{1000} + \frac{1}{500} + \frac{1}{500} + \frac{1}{1000}\right) & -\frac{1}{500} \\ 0 & 0 & -\frac{1}{500} & \left(\frac{1}{50} + \frac{1}{50} + \frac{1}{500}\right) \end{bmatrix} \times \begin{bmatrix} V_1 \\ V_2 \\ V_3 \\ V_4 \end{bmatrix}$$

CONTINUES

2. CONTINUED

$i_1 = \frac{V_4 - V_3}{500}$, $i_2 = \frac{V_3}{500}$, so the left-hand matrix is modified and all rows are multiplied by 1000:

$$\begin{bmatrix} -100(V_4 - V_3) \\ 100V_3 \\ 100V_4 - 200V_3 \\ 20V_s \end{bmatrix} = \begin{bmatrix} 2 & 0 & -1 & 0 \\ 0 & 2 & -1 & 0 \\ -1 & -1 & 7 & -2 \\ 0 & 0 & -2 & 42 \end{bmatrix} \times \begin{bmatrix} V_1 \\ V_2 \\ V_3 \\ V_4 \end{bmatrix}$$

V_3 coefficients from the left-hand matrix subtract from 3rd column values; V_4 coefficients subtract from 4th column values

$$\begin{bmatrix} 0 \\ 0 \\ 0 \\ 20V_s \end{bmatrix} = \begin{bmatrix} 2 & 0 & (-1-100) & (0+100) \\ 0 & 2 & (-1-100) & 0 \\ -1 & -1 & (7+200) & (-2-100) \\ 0 & 0 & -2 & 42 \end{bmatrix} \times \begin{bmatrix} V_1 \\ V_2 \\ V_3 \\ V_4 \end{bmatrix} = \begin{bmatrix} 2 & 0 & -101 & 100 \\ 0 & 2 & -101 & 0 \\ -1 & -1 & 207 & -102 \\ 0 & 0 & -2 & 42 \end{bmatrix} \times \begin{bmatrix} V_1 \\ V_2 \\ V_3 \\ V_4 \end{bmatrix}$$

3. first convert the controlled current source to the thevenin equivalent:

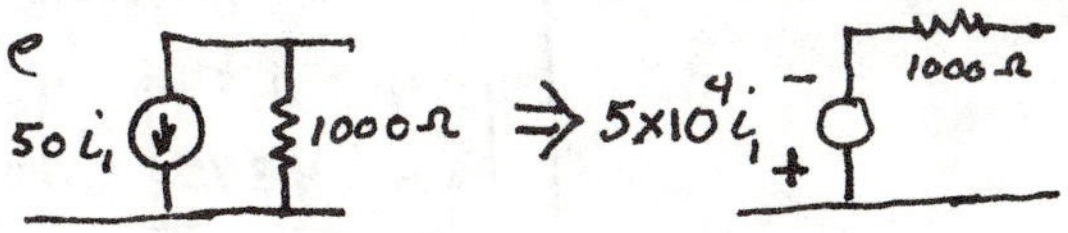

next reflect the 200Ω Load to the primary circuit

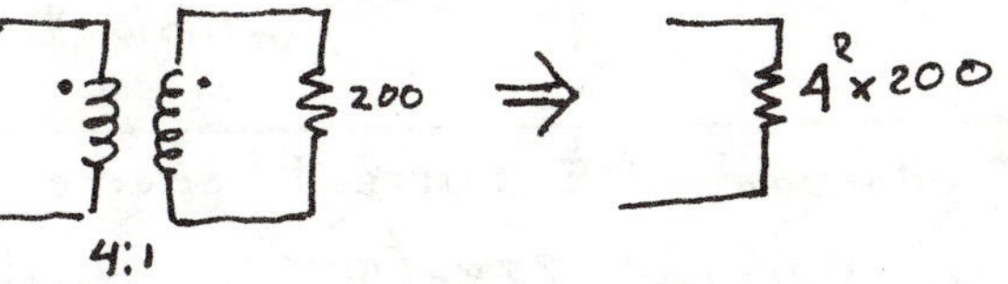

INTERMEDIATE CIRCUIT

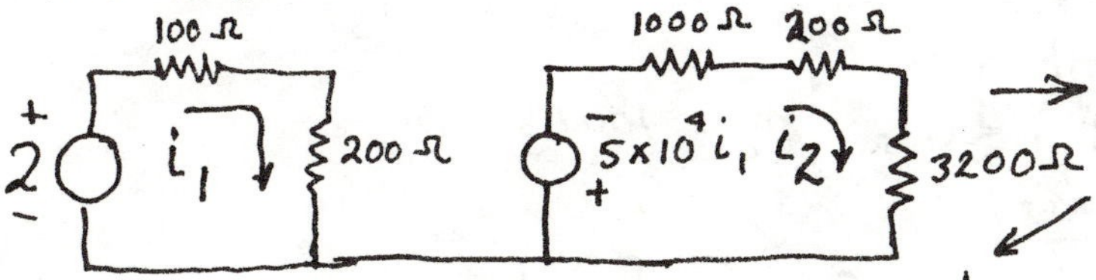

$$\begin{bmatrix} 2 \\ -5\times10^4 i_1 \end{bmatrix} = \begin{bmatrix} 300 & 0 \\ 0 & 3400 \end{bmatrix} \times \begin{bmatrix} i_1 \\ i_2 \end{bmatrix}$$

left-hand matrix coef. of i_1 subtracted from first column of same row: →

$$\begin{bmatrix} 2 \\ 0 \end{bmatrix} = \begin{bmatrix} 300 & 0 \\ 5\times10^4 & 3400 \end{bmatrix} \times \begin{bmatrix} i_1 \\ i_2 \end{bmatrix}$$

$$\Delta = \begin{vmatrix} 300 & 0 \\ 5\times10^4 & 3400 \end{vmatrix} = 1.02\times10^6 \qquad i_2 = \frac{\begin{vmatrix} 300 & 2 \\ 5\times10^4 & 0 \end{vmatrix}}{\Delta} = \frac{-10^5}{1.02\times10^6} = -98 \text{ mA}$$

IDEAL TRANSFORMER: $N_p i_p + N_s i_s = 0$ $\qquad i_{200} = -4 i_2 = 392$ mA ↑

4.

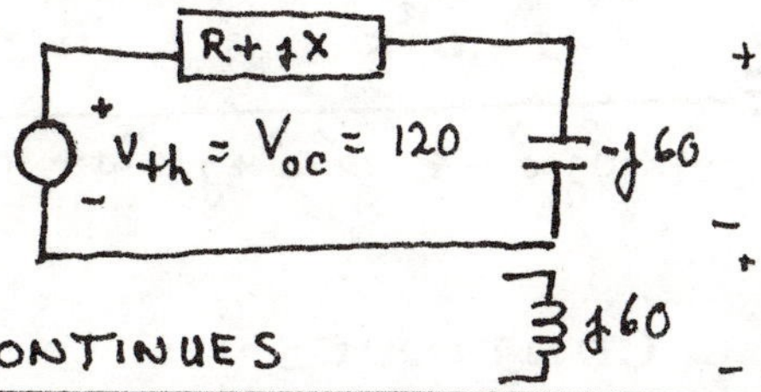

$$360 = \frac{120\,|-j60|}{\sqrt{R^2 + (X-60)^2}} \quad \therefore R^2 + X^2 - 120X + 60^2 = 400$$

$$51.43 = \frac{120\,|j60|}{\sqrt{R^2 + (X+60)^2}} \quad \therefore R^2 + X^2 + 120X + 60^2 = 19598.9$$

CONTINUES

4 CONTINUED

Subtract the first equation from the second and obtain that: $240x = 19198.912$ $x = 79.995$

from first eqn: $R^2 + (19.995)^2 = 400$ $R = .447\,\Omega$

Then for 120Ω capacitance: $|V_c| = \dfrac{120\,|-j120|}{\sqrt{.447^2 + (79.995 - 120)^2}} = 359.9$

5. $R_2 = 1.5K$ @ 20°C $R_1 = 1.5K(1 - .04(20-25)) = 1.8K$ @ 30°C $R_1 = 1.5K(1 - .04(30-25)) = 1.2K$

Kirchhoff current law equations: V_1 on left, V_2 on right

$$\frac{24 - V_1}{R_1} = \frac{V_1}{1K} + \frac{V_1 - V_2}{.8K} \;:\; 24 = \left(1 + \frac{2.25 R_1}{K}\right)V_1 - \frac{1.25 R_1}{K} V_2 = \begin{cases} 5.05V_1 - 2.25V_2 & @\,20^\circ C \\ 3.7V_1 - 1.5V_2 & @\,30^\circ C \end{cases}$$

$$\frac{24 - V_2}{1.5K} = \frac{V_2}{1K} + \frac{V_2 - V_1}{.8K} \;:\; 24 = -1.875V_1 + 4.375V_2$$

$$I_m = \frac{V_1 - V_2}{.8K} = \begin{cases} \dfrac{\begin{vmatrix} 24 & -2.25 \\ 24 & 4.375 \end{vmatrix} - \begin{vmatrix} 5.05 & 24 \\ -1.875 & 24 \end{vmatrix}}{.8K \begin{vmatrix} 5.05 & -2.25 \\ -1.875 & 4.375 \end{vmatrix}} = -.504 \text{ mA @ } 20^\circ C \\[2em] \dfrac{\begin{vmatrix} 24 & -1.5 \\ 24 & 4.375 \end{vmatrix} - \begin{vmatrix} 3.7 & 24 \\ -1.875 & 24 \end{vmatrix}}{.8K \begin{vmatrix} 2.7 & -1.5 \\ -1.875 & 4.375 \end{vmatrix}} = +.673 \text{ mA @ } 30^\circ C \end{cases}$$

6. Convert KE current source ∥ 5K to thevenin equivalent & write loop current equations:

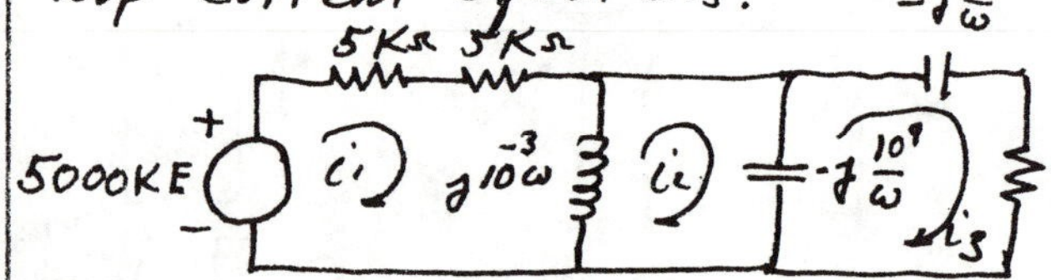

$e_o = 400\,i_3$

$$\begin{bmatrix} 5000\,KE \\ 0 \\ 0 \end{bmatrix} = \begin{bmatrix} (10^4 + j10^{-3}\omega) & -j10^{-3}\omega & 0 \\ -j10^{-3}\omega & (j10^{-3}\omega - j\frac{10^9}{\omega}) & j\frac{10^9}{\omega} \\ 0 & j\frac{10^9}{\omega} & (400 - j\frac{2\times10^9}{\omega}) \end{bmatrix} \times \begin{bmatrix} i_1 \\ i_2 \\ i_3 \end{bmatrix}$$

then

$$400\,i_3 = \frac{\begin{vmatrix} (10^4 + j10^{-3}\omega) & -j10^{-3}\omega & 5000KE \\ -j10^{-3}\omega & (j10^{-3}\omega - j\frac{10^9}{\omega}) & 0 \\ 0 & j\frac{10^9}{\omega} & 0 \end{vmatrix} \times 400}{\begin{vmatrix} 10^4 + j10^{-3}\omega & -j10^{-3}\omega & 0 \\ -j10^{-3}\omega & (j10^{-3}\omega - j\frac{10^9}{\omega}) & j\frac{10^9}{\omega} \\ 0 & j\frac{10^9}{\omega} & (400 - j\frac{2\times10^9}{\omega}) \end{vmatrix}} \cong \frac{-2\times10^{12}KE}{-4\times10^{3}j\omega^3 - 2\times10^{10}\omega^2 + j10^{15}\omega + 10^{22}}$$

CONTINUES

6 CONTINUED

IN ORDER That $e_o = E$, it is necessary that $-4\times10^{3}j\omega^{3} + j10^{15}\omega = 0$

so that $\omega^2 = \frac{10^{12}}{4}$ [second answer] $\omega = .5\times10^{6}$

then
$$\frac{-2\times10^{12}KE}{-2\times10^{10}\left(\frac{10^{12}}{4}\right) + 10^{22}} = E \quad \text{or } K = .25\times10^{10}$$

1. $\int_{\omega t=\theta}^{\omega t=\theta+2\pi} \sin(\omega t+\alpha)\,dt =$

$$= -\frac{1}{\omega}\cos(\omega t+\alpha)\Big|_{\omega t=\theta}^{\omega t=\theta+2\pi}$$

$$= -\frac{1}{\omega}\left[\cos(\theta+\alpha+2\pi) - \cos(\theta+\alpha)\right]$$

$$= -\frac{1}{\omega}\left[\cos(\theta+\alpha) - \cos(\theta+\alpha)\right] \equiv 0$$

2.

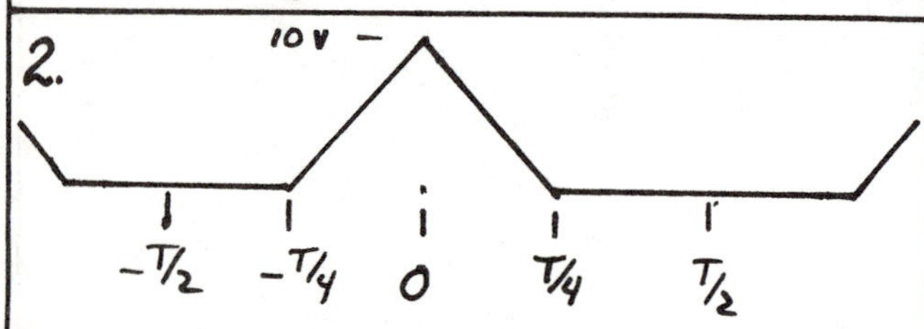

for $-T/2 < t < T/2$ being a cycle from eq. 3.2

POSITIVE AREA $= \frac{1}{2}\left(10 \times \frac{T}{2}\right) = 2.5T$

Negative AREA $= 0$

$$V_{AVG} = \frac{2.5T}{T} = 2.5 \text{ volts} \quad (a)$$

$$V_{RMS}^2 = \frac{1}{T}\int_{-T/2}^{T/2} v(t)^2\,dt \qquad (\text{eq. } 3.15)$$

$$v(t) = \begin{cases} 0 & \text{for } -T/2 < t < -T/4 \\ 10 + \frac{40t}{T} & \text{for } -T/4 < t < 0 \\ 10 - \frac{40t}{T} & \text{for } 0 < t < T/4 \\ 0 & \text{for } T/4 < t < T/2 \end{cases}$$

$$V_{RMS}^2 = \frac{1}{T}\int_{-T/4}^{0}\left(100 + \frac{800}{T}t + \frac{1600}{T^2}t^2\right)dt + \frac{1}{T}\int_{0}^{T/4}\left(100 - \frac{800}{T}t + \frac{1600}{T^2}t^2\right)dt$$

$$V_{RMS}^2 = \frac{1}{T}\left[100t + \frac{400}{T}t^2 + \frac{1600}{3T^2}t^3\right]_{-T/4}^{0} + \frac{1}{T}\left[100t - \frac{400}{T}t^2 + \frac{1600}{3T^2}t^3\right]_{0}^{T/4}$$

$$= 25 - 25 + \frac{25}{3} + 25 - 25 + \frac{25}{3}$$

$$= 16.667$$

$$V_{RMS} = 4.08 \quad (b)$$

3. $V_{AVG} = \dfrac{10 \times 2V \times 1mS}{50\,mS} = 0.4$ volts

$$V_{RMS}^2 = \frac{10 \times (2V)^2 \times 1mS}{50\,mS} = 0.8 \text{ Volts}^2$$

$V_{RMS} = .894$ volts

4. for full scale of 9999, the resolution is 0.5. thus the precision is

$$\frac{0.5}{9999} \times 100 = .0005$$

5. peak at $t = .001$

$\therefore \cos(\omega[t - .001])$

positive zero crossing at $t = -.0015$

$\therefore \sin(\omega[t + .0015])$

$$\omega = \frac{2\pi}{T} = 200\pi$$

$\therefore \cos(200\pi t - .2\pi)$; $\sin(200\pi t + .3\pi)$

6.

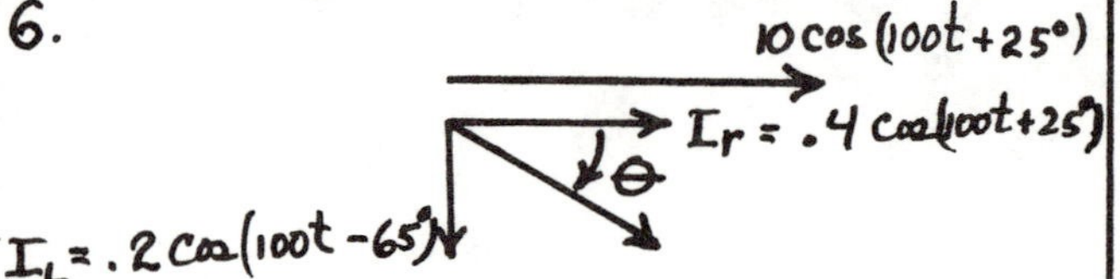

$$|I|_{total} = \sqrt{.2^2 + .4^2} = .447$$

$$\theta = -\tan^{-1}\frac{.2}{.4} = -26.56°$$

$$I_{total} = .447\cos(100t + 25° - 26.56°)$$

$$= .447\cos(100t - 1.56°)$$

$$I_{RMS} = \frac{.447}{\sqrt{2}} = .316 \text{ A}$$

7. $S = VI^* = \frac{10}{\sqrt{2}}\angle 25^\circ \times \frac{.447}{\sqrt{2}}\angle 1.56^\circ = 2 + j1$

$S = \sqrt{5}\angle 26.6^\circ$

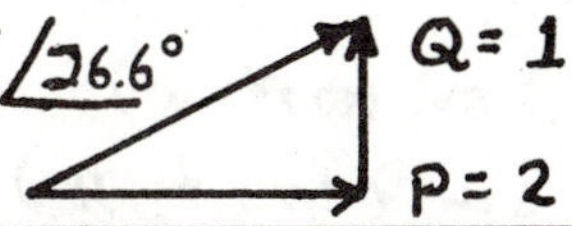

8. $\frac{|V|^2}{|Z_c|} = Q \therefore 100C = \frac{1}{\left(\frac{10}{\sqrt{2}}\right)^2} = .02$

$C = .0002\text{ F} = 200\ \mu\text{F}$

CONCENTRATES

1. For 3V scale $R_3 = \frac{3}{.001} = 3000$

" 10 V " $R_{10} = \frac{10}{.001} = 10{,}000$

" 30 V " $R_{30} = \frac{30}{.001} = 30{,}000$

" 100 V " $R_{100} = \frac{100}{.001} = 100{,}000$

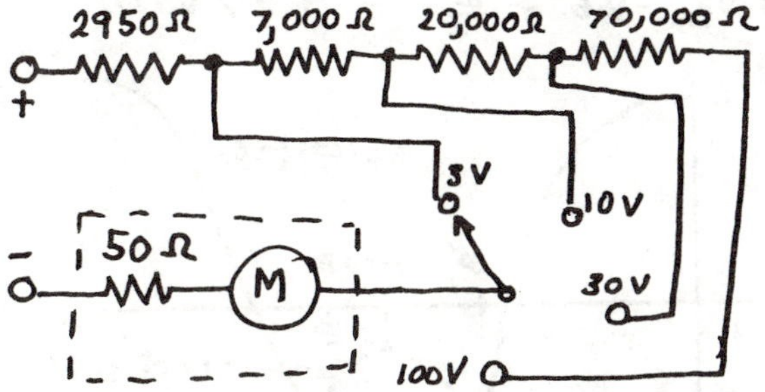

2.

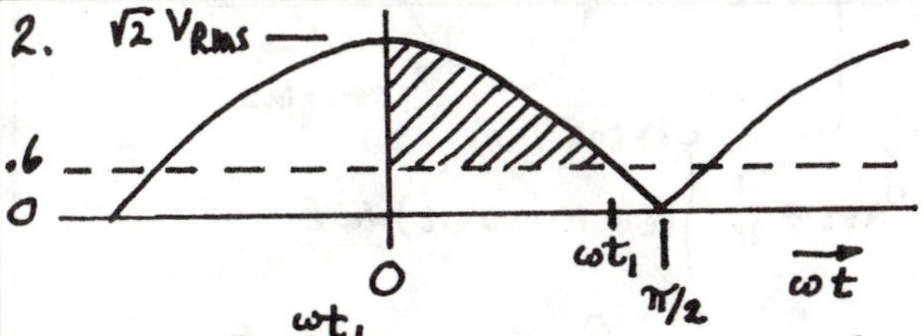

$V_{avg} = \frac{2}{\pi}\int_0^{\omega t_1}(\sqrt{2}\,V_{rms}\cos\omega t - .6)\,d\omega t$

where $\omega t_1 = \cos^{-1}\frac{.6}{\sqrt{2}\,V_{rms}}$

$V_{avg} = \frac{2}{\pi}\left[\sqrt{2}\,V_{rms}\sin\omega t_1 - .6\,\omega t_1\right]$

full-scale voltage for the meter is $.125 \times 10^{-3} \times 10^6 = 125$ volts

solving for each V_{avg}, taking into account a deflection of $50^\circ \times \frac{V_{AVG}}{125}$:

V_{rms}:	10	20	30	40	50
V_{AVG}:	8.41	17.41	26.41	35.41	44.42
Angle:	3.36°	6.96°	10.56°	14.17°	17.77°

V_{rms}:	60	70	80	90	100
V_{AVG}:	53.42	62.42	71.42	80.43	89.43
Angle:	21.37°	24.97°	28.57°	32.17°	35.77°

3. The signal power is represented in units of volts² (squared).

$P_{signal} = \frac{V_{in}^2}{R} = \frac{1}{R}\frac{1}{T}\int_{-T/2}^{T/2} f(t)^2\,dt$

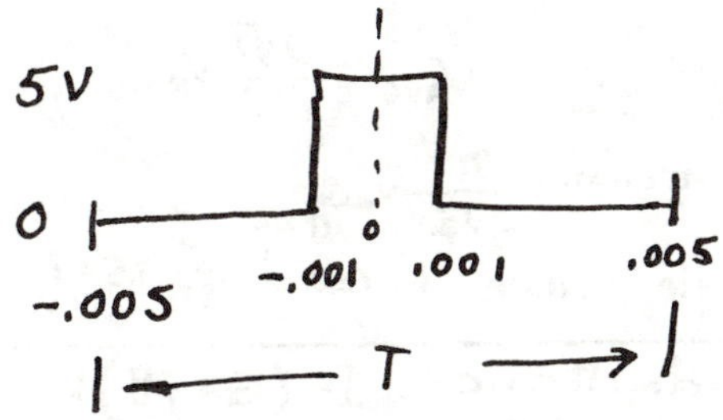

$P_{signal} = \frac{5^2 \times .002}{.001R} = \frac{5}{R}$

The output voltage is represented by the truncated fourier series: $\frac{a_0}{2}, \frac{a_1}{2}, \ldots$ up to 500Hz

the fundamental is $\frac{1}{.01} = 100$ Hz

So the terms up to 500 Hz are necessary. Using Cosine Series due to even symmetry:

$a_n = \frac{2}{.01}\int_{-.001}^{.001} 5\cos 200\pi n t\,dt$

$= \frac{5}{\pi n}\sin\frac{2\pi n}{5}$

$\frac{a_0}{2} = \frac{5 \times .002}{.01} = 1$

$a_1 = 1.51$ $a_4 = -.38$

$a_2 = .47$ $a_5 = 0.00\ldots$

$a_3 = -.31$

CONCENTRATES

3. CONTINUED

the output power is:

$$P_{out} = \frac{1}{R}\left[\frac{a_0^2}{2} + \frac{a_1^2}{2} + \frac{a_2^2}{2} + \frac{a_3^2}{2} + \frac{a_4^2}{2} + \frac{a_5^2}{2} \right]$$

$$= \frac{1}{R}\left[1^2 + \frac{1.51^2}{2} + \frac{.47^2}{2} + \frac{.31^2}{2} + \frac{.38^2}{2} + 0 \right]$$

$$= \frac{1}{R}\, 2.38$$

$$\frac{P_{out}}{P_{sig}} = \frac{2.38}{5.00} \times 100 = 48\%$$

4.

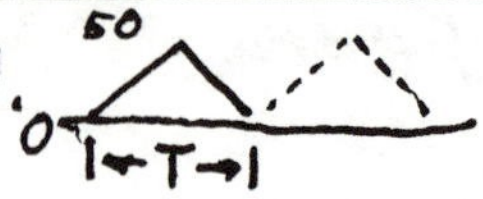

$$V_{AVG} = \frac{50 \times (T/2)}{T} = 25$$

$$V_{AVG} = \frac{2\sqrt{2}}{\pi} V_{RMS}$$

so meter reads $\frac{\pi}{2\sqrt{2}} V_{avg}$

for triangle wave it reads $\frac{\pi}{2\sqrt{2}} 25 = 27.77$ volts

5. use admittance: $I = (G + jB)V$

$$-30° = \tan^{-1}\frac{B}{G}\ ;\ G = \frac{200\text{W}}{(100\text{ V})^2} = .0200$$

$$B = -.0200 \tan 30° = -.0115$$

$$B_{comp} = +.0115 \quad X_{comp} = -86.6$$

6. (a) $X_C = -\frac{1}{377\,C} = -29.97$

$$X_L = 377L = 25.00$$

(b) $G = \frac{1}{R} = 0.1$; $B_C = \frac{-1}{X_C} = .033$; $B_L = \frac{-1}{X_L} = -.04$

(c) $I_R = .1 \times 230 = 23$

$$I_C = j.033 \times 230 = j\,7.7$$

$$I_L = -j.04 \times 230 = -j\,9.2$$

$$I_t = 23 - j1.5$$

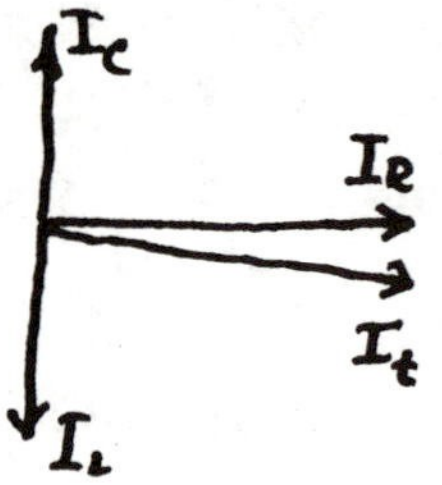

(d)

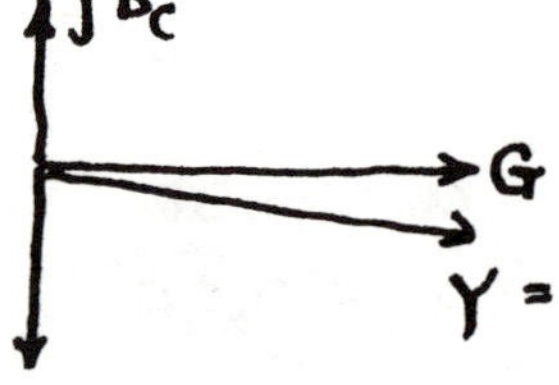

$$Y = .1 - j.007$$

(e) $I = YV = (1.002\angle{-4°}) \times 230\angle{0°}$ (RMS)

$$= 23.1\angle{-4°}$$

$$i(t) = \sqrt{2} \times 23.1 \sin(377t - 4°)$$

$$= 32.6 \sin(377t - 4°)$$

circuit is inductive as voltage leads current.

(f) $\text{p.f.} = \cos(-4°) = .998$

(g) $S = P + jQ = VI^* = 230\angle{0°} \times 23.1\angle{4°}$

$$= 5336\angle{4°} = 5323 + j\,372$$

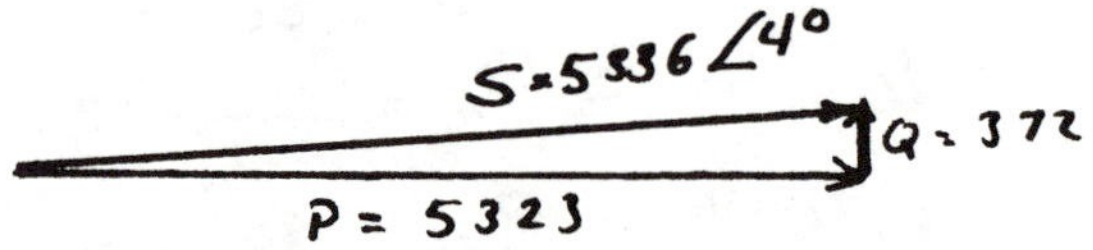

TIMED

1.(a) $V_{RMS} = \sqrt{6^2 + \frac{10^2}{2}} = 9.27$ volts

(b)

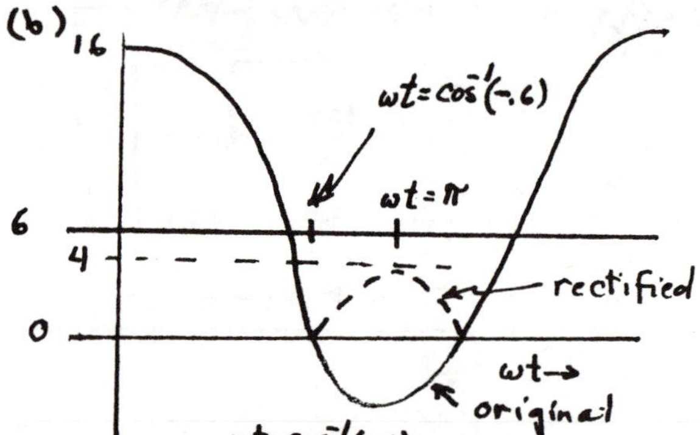

(c)

$$V_{AVG} = \frac{1}{\pi}\int_0^{\omega t = \cos^{-1}(-.6)} (6 + 10\cos\omega t)\,d\omega t + \frac{1}{\pi}\int_{\omega t = \cos^{-1}(-.6)}^{\pi} -(6 + 10\cos\omega t)\,d\omega t$$

$$= \frac{1}{\pi}\Big[6\omega t + 10\sin(\cos^{-1} -.6)\Big]_{\omega t = \cos^{-1} -.6}$$

$$- \frac{1}{\pi}\Big[6\pi + 10\sin\pi - 6\cos^{-1}(-.6) - 10\sin(\cos^{-1} -.6)\Big]$$

$$= \frac{1}{\pi}\Big[12(\cos^{-1} -.6) - 6\pi + 20\sin(\cos^{-1} -.6)\Big]$$

$$= 8.485 - 6 + 5.093 = 7.55$$

meter reads

$$\frac{\pi}{2\sqrt{2}} V_{AVG} = 8.39 \text{ volts}$$

TIMED

2.

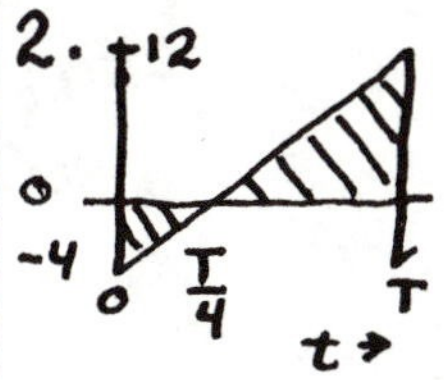

(a) $V_{AVG} = \frac{1}{2}\frac{\frac{3T}{4}\times 12 - \frac{T}{4}\times 4}{T}$

$= 4$ volts

(b) Assuming a half-wave rectifier which responds to the positive voltages, it is calibrated to read $V_{RMS} = \frac{\pi}{\sqrt{2}} V_{AVG}$

positive only: $V_{AVG} = \frac{1}{2}\frac{\frac{3T}{4}\times 12}{T} = 4.5\text{V}$

READING: $\frac{\pi}{\sqrt{2}} \times 4.5 = 10.10$ volts

(c) Meter calibrated: $V_{RMS} = \frac{V_{P-P}}{2\sqrt{2}}$

$V_{P-P} = 12 - (-4) = 16$

READING: $\frac{16}{2\sqrt{2}} = 5.66$ volts

(d) TRUE RMS reading:

$V_{RMS}^2 = \frac{1}{T}\int_0^T v(t)^2\,dt \quad : v(t) = 16\frac{t}{T} - 4$

$V_{RMS}^2 = \frac{1}{T}\int_0^T \left(256\frac{t^2}{T^2} - 128\frac{t}{T} + 16\right)dt = 37.33$

$V_{RMS} = 6.11$ volts

3. $R_{series} = R_a + 50$

$R_{shunt} = R_1 + R_2 + R_3 + R_4 + R_5 \times 1000$

$R_{TOTAL} = R_{series} + R_{shunt}$

In 1.5 mA position, $I_A = 1\text{ mA}$ when $I_{in} = 1.5\text{ mA}$. Using current division:

$1\text{ mA} = 1.5\text{ mA} \times \frac{R_{shunt}}{R_{TOTAL}}$ ∴ $R_{series} = 500\,\Omega$

$R_a = 450\,\Omega$

$R_{TOTAL} = 1500$

in 15 mA position

$1\text{ mA} = 15\text{ mA} \times \frac{R_2 + R_3 + R_4 + R_5}{1500}$

∴ $R_2 + R_3 + R_4 + R_5 = 100$

∴ $R_1 = 900\,\Omega$

in 150 mA position:

$1\text{ mA} = 150\text{ mA} \times \frac{R_3 + R_4 + R_5}{1500}$; $R_3 + R_4 + R_5 = 10$

∴ $R_2 = 90\,\Omega$

in 1.5 A position

$1\text{ mA} = 1500\text{ mA} \times \frac{R_4 + R_5}{1500}$; $R_4 + R_5 = 1\,\Omega$

∴ $R_3 = 9\,\Omega$

in 15 A position

$1\text{ mA} = 15{,}000\text{ mA} \times \frac{R_5}{1500}$; $R_5 = .1\,\Omega$

∴ $R_4 = .9\,\Omega$

4. $V_{AVG} = 5\text{ V}$ from d'Arsonval meter

$V_{max} - V_{min} = 2\sqrt{2}\,V_{RMS} = 19.8$ volts

$V_{AVG} = \frac{V_{MAX} + V_{min}}{2} = 5$

$V_{max} + V_{min} = 10$

$V_{max} - V_{min} = 19.8$

$V_{max} = 14.9\text{ V}$

$V_{min} = 4.9\text{ V}$

5. $P = V|I|\cos\theta$ ∴ $|I| = \frac{P}{\text{p.f.} \times V} = \frac{10^5}{.867 \times 1320} = 87.4\text{ A}$

$I = 87.4\angle -\cos^{-1}(.867) = 87.4\angle -29.9^\circ$

$S = VI^* = 100\text{ KW} + j\,57.5\text{ KVAR}$

(a) for p.f. = .895 lagging

$Q = P\tan(\cos^{-1}.895) = 49.8\text{ KVAR}$

$Q_{correcting} = 49.8 - 57.5 = -7.7\text{ KVAR}$

$Q_c = \frac{V^2}{X_c} \quad X_c = \frac{V^2}{Q_c} = \frac{1320^2}{-7700} = -226\,\Omega$ @ 1.32 KV

(b) for p.f. = .95 leading

$Q = -P\tan(\cos^{-1}.95) = -32.9\text{ KVAR}$

$Q_{correcting} = -32.9 - 57.5 = -90.4\text{ KVAR}$

$X_c = \frac{1320^2}{90400} = 19.3\,\Omega$ @ 1.32 KV

1.(a) KVL (Kirchhoff's voltage law)

$15 = 4\,i_{4\Omega} + v_L$

$v_L = L\frac{di_L}{dt} \to 0$ in steady state

with $v_L = 0$ $\quad i_{4\Omega} = i_L = \frac{15}{4}$ A

(b) KVL: $8i_L + .1\frac{di_L}{dt} = 0$

but $\frac{di_L}{dt} \to 0$, so $i_L = 0$

2.(a) After a long time $\frac{dv_c}{dt} \to 0$

as $i_c = C\frac{dv_c}{dt}$, $i_c \to 0$

then $v_c = \frac{8}{4+8} \times 15 = 10$ V

(b) as $\frac{dv_c}{dt} \to 0$, and $i_c = 0$

as $v_c = -8i_c$, $v_c = 0$

3. The thevenin equivalent circuit is $v_{th} = 10\cos(10t + 20^\circ)$ and $Z_{th} = \frac{8}{3}$ ohms, as seen from L, for the switch closed. This a sinusoidal analysis problem.

$Z_L = j\omega L = j\,10 \times .1 = j1$

$I_L = \frac{V_{th}}{Z_{th} + Z_L} = \frac{10\angle 20^\circ}{8/3 + j1} = 3.51\angle -.56^\circ$

(a) $i_L(t) = 3.51\cos(10t - .56^\circ)$

(b) for switch open, steady state current is zero.

4. From 2(a) $v_c(0) = 10$

KVL: $v_c + 8i_c = 0$; $i_c = 10^{-6}\frac{dv_c}{dt}$

$8 \times 10^{-6}\frac{dv_c}{dt} + v_c = 0$: $v_c = Ae^{-t/\tau}$

$\tau = 8 \times 10^{-6}$; at $t=0$ $v_c = 10 = A$

then $v_c(t) = 10e^{-1.25 \times 10^5 t}$ for $t \geq 0$

5. from 2(b) $v_c(0) = 0$

KVL: $10 = \frac{8}{3}i_c + v_c$; $i_c = 10^{-6}\frac{dv_c}{dt}$

$10 = \frac{8}{3} \times 10^{-6}\frac{dv_c}{dt} + v_c$

Solution: $v_c = A + Be^{-t/\tau}$; $\tau = \frac{8}{3} \times 10^{-6}$

as $t \to \infty$ (from 2a) $v_c \to 10$

$A + Be^{-\infty} = A = 10 = v_c(\infty)$

at $t=0$ $v_c = 0 = A + Be^{0} = A + B = 0$

$B = -A = -10$

$v_c(t) = 10(1 - e^{-3.75 \times 10^5 t})$ for $t \geq 0$

6 $b^2 - 4ac = 9$: $s = \frac{-5 \pm 3}{2} = -4, -1$

$v(t) = Ae^{-4t} + Be^{-t} + v_{ss}$

$v_{ss} = C\cos 2t + D\sin 2t$

$\frac{dv_{ss}}{dt} = -2C\sin 2t + 2D\cos 2t$

$\frac{d^2v_{ss}}{dt^2} = -4C\cos 2t - 4D\sin 2t$

in original equation

$12\sin 2t = (-4C + 10D + 4C)\cos 2t + (-4D - 10C + 4D)\sin 2t$

then $v_{ss} = -1.2\cos 2t$

$\frac{dv_{ss}}{dt} = 2.4\sin 2t$

$v(0) = -5 = A + B - 1.2$

$\frac{dv}{dt}(0) = 2 = -4A - B + 0$

$\therefore A = .6$, $B = -4.4$

$v(t) = .6e^{-4t} - 4.4e^{-t} - 1.2\sin 2t$

7. Transform circuit using impedances

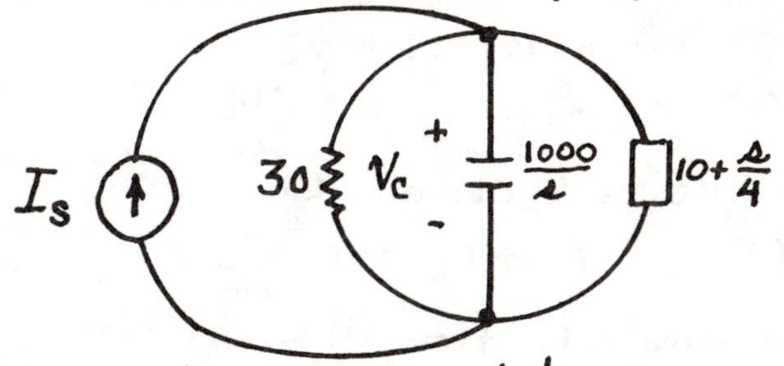

use Kirchoff's current law:

$$I_s = V_C\left[\frac{1}{30} + \frac{s}{1000} + \frac{1}{10+\frac{s}{4}}\right]$$

solve for V_c:

$$G(s) = \frac{V_c}{I_s} = \frac{3000\left(10+\frac{s}{4}\right)}{\frac{3}{4}s^2 + 55s + 4000}$$

8. $Z = 10 + .5s + \frac{10^4}{s}$

$$= .5\left[\frac{s^2 + 20s + (2\times10^4)}{s}\right]$$

matching:

$$s^2 + \frac{\omega_o}{Q}s + \omega_o^2 = s^2 + 20s + (2\times10^4)$$

then $\omega_o = 100\sqrt{2}$, $Q = 5\sqrt{2}$

$$BW = \frac{\omega_o}{Q} = 20 \text{ r/s}$$

9. $Y = .1 + 10^{-4}s + \frac{2}{s}$

$$= \frac{10^{-4}}{s}\left[s^2 + 10^3 s + 2\times10^4\right]$$

then $\omega_o = 100\sqrt{2}$, $Q = \frac{\sqrt{2}}{10}$

$$BW = \frac{\omega_o}{Q} = 1000 \text{ rad/sec}$$

10. With the switch open for a long time, the right capacitor has charged to 10 volts, and the left is discharged to zero volts. At the instant of switching, the capacitors can be modeled as constant voltage sources which have the voltages of the capacitances just prior to switching. The t=0 equivalent circuit is:

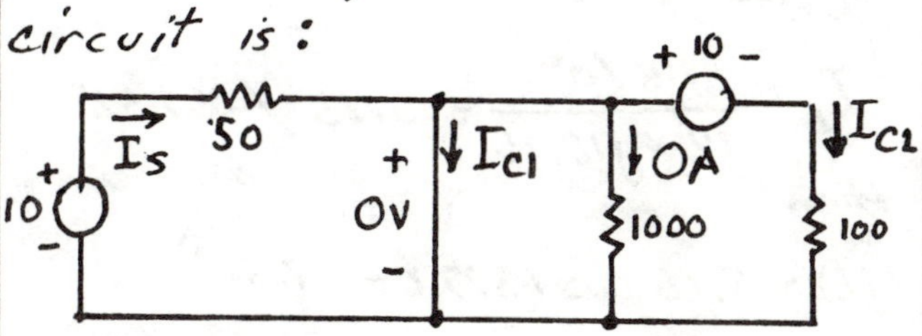

$I_s = \frac{10}{50} = .2\text{A}$, $I_{C2} = -\frac{10}{100} = -.1\text{A}$

$$I_{C1} = I_s - I_{C2} = .3\text{A}$$

CONCENTRATES

1. $b^2 - 4ac = 16 - 16 = 0 \quad \therefore s = \frac{-4}{2} = -2$

$$i(t) = Ae^{-2t} + Bte^{-2t} + I_{ss}$$

$$I_{ss} = C\cos 4t + D\sin 4t$$

steady state analysis

$$20\sin 4t = -16C\cos 4t - 16D\sin 4t$$
$$-16C\sin 4t + 16D\cos 4t$$
$$+4C\cos 4t + 4D\sin 4t$$

$$= (-12C + 16D)\cos 4t + (-12D - 16C)\sin 4t$$

matching coefficients of sine & cosine:

$$-12C + 16D = 0\ ;\ -12D - 16C = 20$$

solving for C and D:

$$C = -.8,\ D = -.6$$

$$i(t) = Ae^{-2t} + Bte^{-2t} - .8\cos 4t - .6\sin 4t$$

$$i(0) = 0 = A - .8 \quad \therefore A = .8$$

$$\frac{di(t)}{dt} = -2Ae^{-2t} + Be^{-2t} - 2Bte^{-2t} + 3.2\sin 4t - 2.4\cos 4t$$

$$\frac{di(0)}{dt} = -2A + B - 2.4 = 4 \quad \therefore B = 8$$

$$i(t) = .8e^{-2t} + 8te^{-2t} - .8\cos 4t - .6\sin 4t$$

$$= .8e^{-2t} + 8te^{-2t} - \cos(4t - 36.9°)$$

2. The circuit prior to switching is analyzed (phasor) to obtain i_L at $t=0$:

$$I_L = \frac{115\angle 0^\circ}{12+j18.75} = 5.15\angle -57.52^\circ$$

then

$$i_L(t) = 5.15\cos(377t - 57.52^\circ)$$

$$i_L(0) = 5.15\cos(-57.52^\circ) = 2.76\ A$$

which is the inductance current at switching.

Because no information has been given, assume capacitance is initially uncharged ($v_c(0)=0$). Replace capacitance by a short circuit (zero voltage source) and inductance with constant current source (2.76 A), and voltage source by value at $t=0$ to obtain initial condition circuit.

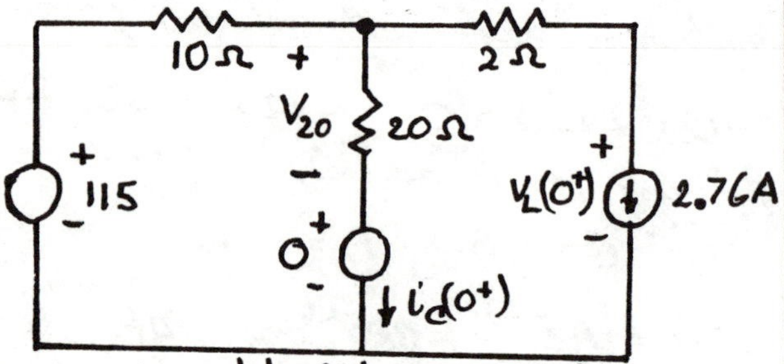

Write Kirchhoff's current law for V_{20}:

$$\frac{115 - V_{20}}{10} = \frac{V_{20}}{20} + 2.76 \therefore V_{20} = 58.24$$

then

$$i_c(0^+) = \frac{V_{20}}{20} = 2.91\ A$$

$$V_L(0^+) = V_{20} - (2.76 \times 2)$$

$$= 52.72\ V$$

3. Kirchhoff's voltage law (twice)

(1) $100 = 100\, i_1 + v_C$

(2) $v_C = 10\, i_L + 1\dfrac{di_L}{dt}$

Kirchhoff's current law

(3) $i_1 = i_C + i_L = 10^{-4}\dfrac{dv_C}{dt} + i_L$

eliminate i_1 from (1) and (3)

(4) $100 = 100\, i_L + 10^{-2}\dfrac{dv_C}{dt} + v_C$

Laplace transform (2) and (4)

[2] $V_C = (10+s)\, I_L - i_L(0^+)$

$= (s+10)I_L - 1$

[4] $\dfrac{100}{s} = 100\, I_L + 10^{-2} s V_C - 10^{-2} v_C(0^+) + V_C$

$\dfrac{100}{s} = 100\, I_L + \dfrac{s+100}{100} V_C - .2$

Solve for V_C from [2] and [4]

$$V_C(s) = \frac{20s^2 + 200s + 10^5}{s(s^2 + 110s + 1.1\times 10^4)}$$

this is an underdamped response, so we want it in the form:

$$V_C(s) = \frac{A}{s} + \frac{B(s+55)}{(s+55)^2 + 89.3^2} + \frac{89.3\,C}{(s+55)^2 + 89.3^2}$$

in order to use tabled transform

Solving for A, B and C we obtain $A = 9.09$, $B = 10.91$, $C = -15.68$

then INVERSE TRANSFORMING:

$$v_C(t) = 9.09 + 10.91 e^{-55t}\cos 89.3t - 15.68 e^{-55t}\sin 89.3t$$

$$= 9.09 + 19.10 e^{-55t}\cos[89.3t + 55.2^\circ]$$

4. Initially $W_1 = \frac{1}{2} C_1 V_1^2 = \frac{10^{-6} \times 10^4}{2} = .005$ J

$W_2 = \frac{1}{2} C_2 V_2^2 = 0$

finally $V_1 = V_2 = \frac{Q_1}{C_1} = \frac{Q_2}{C_2}$

but all the charge initially was on C_1, so $Q_1 + Q_2 = Q_{initial}$

$Q_{initial} = 100\, C_1 = 10^{-4}$ Coul.

so $Q_1 + Q_2 = 10^{-4}$ and $Q_2 = \frac{C_2}{C_1} Q_1$

or

$Q_2 (1 + \frac{1}{2}) = 10^{-4}$ $Q_2 = \frac{2}{3} \times 10^{-4}$

$Q_1 = \frac{1}{3} \times 10^{-4}$

and

$V_1 = V_2 = \frac{Q_1}{C_1} = \frac{\frac{1}{3} 10^{-4}}{10^{-6}} = \frac{100}{3}$

so finally

$W_1 = \frac{1}{2} Q_1 V_1 = .000556$ J

$W_2 = \frac{1}{2} Q_2 V_2 = .001666$ J

The difference between the initial and final energy is lost in the resistance in the case where $R \neq 0$:

W_L resistance $= .003333$ J

If $R = 0$, this energy is radiated as an electromagnetic wave.

TIMED

1. The transfer function is obtained by voltage division:

$$\frac{V_{out}}{V_{in}} = \frac{K R_2}{R_1 + R_2} \frac{sCR_1 + 1}{sC \frac{R_1 R_2}{R_1 + R_2} + 1}$$

$$= \frac{K R_2}{R_1 + R_2} \frac{j\omega/z + 1}{j\omega/p + 1}$$

The phase angle is

$$\phi = \tan^{-1} \frac{\omega}{z} - \tan^{-1} \frac{\omega}{p}$$

It can be easily shown that the maximum phase angle occurs at $\omega|_{max\,\phi} = \sqrt{pz}$

i.e. at the geometrical mean of the pole and zero. Then

$$\phi_{max} = \tan^{-1} \sqrt{\frac{p}{z}} - \tan^{-1} \sqrt{\frac{z}{p}}$$

by a trig identity $\left(\tan(A-B) = \frac{\tan A - \tan B}{1 + \tan A \tan B}\right)$

$$\sqrt{\frac{p}{z}} - \sqrt{\frac{z}{p}} = 2 \tan \phi_{max}$$

or

$$\frac{p - z}{\sqrt{pz}} = 2 \tan \phi_{max}$$

then

$$p - z = 2\sqrt{pz} \tan \phi_{max}$$

Let $\sqrt{pz} = \omega_d$, $\phi_{max} = \phi_d$

then $p - z = 2\omega_d \tan \phi_d$

$p = \frac{R_1 + R_2}{C R_1 R_2}$ $z = \frac{1}{CR_1}$

$$p - z = \boxed{\frac{1}{CR_2} = 2\omega_d \tan \phi_d} \quad (a)$$

As

$$\omega_d^2 = pz = \frac{1}{CR_1} \cdot \frac{R_1 + R_2}{C R_1 R_2}$$

using (a) to obtain

$$\boxed{\frac{R_1}{R_2} = \frac{2 \sin \phi_d}{1 - \sin \phi_d}} \quad (b)$$

Considering gain:

$$\left|\frac{V_{out}}{V_{in}}\right|_{\omega = \omega_d} = \frac{K R_2}{R_1 + R_2} \frac{\sqrt{\frac{pz}{z^2} + 1}}{\sqrt{\frac{pz}{p^2} + 1}} = \frac{K R_2}{R_1 + R_2} \sqrt{\frac{p}{z}}$$

but $\sqrt{\frac{p}{z}} = \sqrt{\frac{R_1 + R_2}{R_2}}$

So

$$\left|\frac{V_{out}}{V_{in}}\right|_{\omega = \omega_d} = K \sqrt{\frac{R_2}{R_1 + R_2}} = 1$$

1. Continued

The angle is limited to less than 90°. Because of noise and amplifier stability, the ratio of the pole to the zero is typically limited to a factor of 10. this results in a maximum angle of 54.9°

2. Assume voltages for input and output. Then by voltage division:

$$\frac{V_{out}}{V_{in}} = G(s) = \frac{100}{s(s+1100)} \quad (a)$$

Unit step input: $v_{in}(t) = u(t)$

$V_{in}(s) = \frac{1}{s}$; therefore

$$V_{out}(s) = \frac{100}{s(s+1100)} = \frac{1}{11}\left(\frac{1}{s} - \frac{1}{s+1100}\right)$$

so:

$$v_{out}(t) = .091\, u(t)\left[1 - e^{-1100t}\right] \quad (b)$$

unit pulse input: $v_{in}(t) = u(t) - u(t-.001)$

$$V_{in}(s) = \frac{1}{s} - \frac{e^{-.001s}}{s}$$

$$v_{out}(t) = .091\, u(t)\left[1 - e^{-1100t}\right] - .091\, u(t-.001)\left[1 - e^{-1100(t-.001)}\right]$$

(c)

3. Steady state d.c. $(t = 0^-)$ replace inductance with short replace capacitance with open

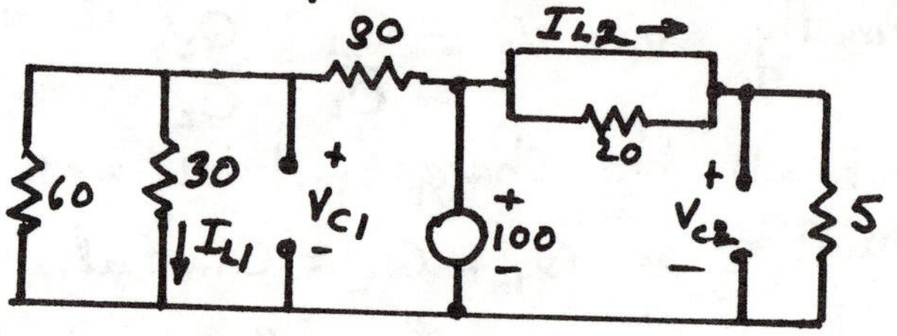

$$I_{L2} = \frac{100}{5} = 20\ A$$

$$V_{C2} = 100\ V$$

$$V_{C1} = \frac{30 \| 60}{30 + 30\|60} \times 100 = 40V$$

$$I_{L1} = \frac{V_{C1}}{30} = \frac{4}{3} A$$

these will remain constant until after $t = 0^+$

at $t = 0^+$, replace inductances with currents at $t = 0^-$, replace capacitances with voltages at $t = 0^-$.

RELEVANT CIRCUIT (redundancy removed)

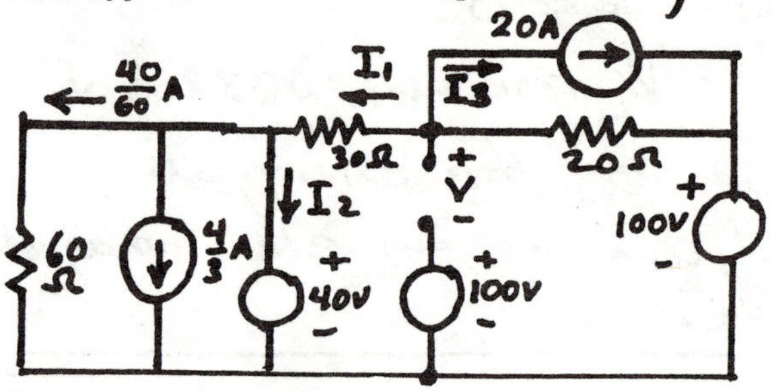

$I_3 = 20A \quad (I_{L2}(0^-))$

KVL right loop

$$40 + 30 I_1 + 20(I_1 + I_3) = 100$$

$$I_1 = -6.8A$$

KCL left of 30Ω

$$I_1 = I_2 + \frac{40}{60} + \frac{4}{3} \qquad I_2 = -8.8A$$

$$V = 40 + 30 I_1 - 100 = -264V$$

$$\frac{dI_3}{dt} = \frac{V}{L} = \frac{-264}{2} = -137\ A/S$$

WARM-UPS

1. $I_{phase} = \frac{S/3}{V} = \frac{2 \times 10^8}{3 \times 230 \times 10^3} = 289.9\,A$

$I_{line} = \sqrt{3}\, I_{phase} = 502\, A$

$I_{g\,(phase)} = I_{phase} \sin(\cos^{-1} .9) = 126.4A$

for unity p.f. $I_{comp} = -I_g$

$X_{comp} = \frac{V}{I_{comp}} = -\frac{230 \times 10^3}{126.4} = -1820\,\Omega$ reactive

2. $G = \frac{P_{oc}}{V_{oc}^2} = \frac{900}{13800^2} = 4.73 \times 10^{-6}$ mhos

$B = \frac{\sqrt{S_{oc}^2 - P_{oc}^2}}{V_{oc}^2} = -1.37 \times 10^{-5}$ mhos react.

$a = \frac{V_{1oc}}{V_{2oc}} = \frac{13800}{460} = 30$

3. $a = I_2 / I_1 = 5$

$P_{sc} = I_1^2 (R_1 + a^2 R_2) = 20^2 (R_1 + a^2 R_2) = 1000$

$R_1 + a^2 R_2 = 2.5$

assume $R_1 = a^2 R_2 = 1.25\,\Omega$, $R_2 = 0.05\,\Omega$

$S = I_1 V_1 = 20 \times 80 = 1600$

$Q = \sqrt{S^2 - P^2} = \sqrt{1600^2 - 1000^2} = 1249$

$Q = I_1^2 (X_1 + a^2 X_2)$: $X_1 + a^2 X_2 = \frac{1249}{20^2} = 3.12\,\Omega$

for $X_1 = a^2 X_2$, $X_1 = 1.56\,\Omega$ reactive

$X_2 = .0624\,\Omega$ reactive

$2\pi f L = X_L$: $L_1 = \frac{1.56}{120\pi} = 4.14 \times 10^{-3}\, H$

$L_2 = \frac{.0624}{120\pi} = .166 \times 10^{-3}\, H$

4. $Z_{base} = \frac{V_{base}}{I_{base}} = \frac{V_{\ell\ell}/\sqrt{3}}{(VA/3)/(V_{\ell\ell}/\sqrt{3})} = \frac{V_{\ell-\ell}^2}{VA}$

$= \frac{440^2}{10{,}000} = 19.36$

$Z = Z_{base} \times Z_{p.u.} = 19.36 \times .8 = 15.49\,\Omega$

$Z_{base\,2} = \frac{440^2}{10^5} = 1.936$

$Z_{p.u.\,2} = \frac{15.49}{1.936} = 8$ p.u.

5. Rated current on the H.V. side is:

$I_{rated} = I_{sc} = \frac{500\, KVA}{115\, KV} = 4.35\,A$

$|Z_{sc}| = \frac{2.5\,KV}{4.35\,A} = 574.7\,\Omega$

$R_{sc} = \frac{P}{I^2} = \frac{435}{4.35^2} = 22.99\,\Omega$

$X_{sc}^2 = |Z_{sc}|^2 - R_{sc}^2$: $X_{sc} = 574.7\,\Omega$

$Z = 22.99 + j574.7 \rightarrow 23 + j575$

$Z_{base} = \frac{V_{rated}^2}{VA_{rated}} = \frac{(115\,KV)^2}{500\,KV} = 26\,450\,\Omega$

$Z_{pu} = \frac{Z}{Z_{base}} = .0009 + j.0217$

6. $V_2 = 1$ (P.U.) at full load

for $I = 1\angle 0^\circ$ $\quad V_1 = 1 + (1\angle 0^\circ)(.014 + j.024)$

$= 1.014 + j.024 = 1.014\angle 1.4^\circ$

half load $I = .5\angle 0$

Assume $V_2 = 1$ $\quad V_1 = 1 + .5(.014 + j.024)$

$= 1.007 + j.012 = 1.007\angle .7^\circ$

but $V_1 = 1.014\angle 1.4^\circ$, so $|V_2| = \frac{1.014}{1.007} = 1.007$

so V_2 (N.L.) $= 1.014$

V_2 (F.L.) $= 1$

V_2 (half load) $= 1.007$

regulation half to full load $= 1.007 - 1 \rightarrow .7\%$

" no load to full load $= 1.014 - 1 \rightarrow 1.4\%$

7. $V_2 = 1$ (p.u.) at full load.

full load $I = 1\angle(\cos^{-1} .4) = 1\angle -66.4^\circ$

$V_1 = 1 + (1\angle 66.4^\circ)(.014 + j.024) = 1.0276\angle 2^\circ$

half load

Assume $V_2 = 1$: $V_1 = 1 + (.5\angle 66.4^\circ)(.014 + j.024)$

$= 1.0139\angle -6.7^\circ$

but $|V_1| = 1.0276$ $\quad \therefore |V_2| = \frac{1.0276}{1.0139} = 1.0134$

then

Regulation half to full load $= 1.0134 - 1 \rightarrow 1.34\%$

" no load to " " $= 1.0276 - 1 \rightarrow 2.76\%$

8. $I = 1\angle -\cos^{-1}.8 = 1\angle -36.9^\circ$ $V_2 = 1$ p.u.

$V_1 = V_2 + (1\angle -36.9^\circ)(.03 + j.04) = 1.048\angle .77^\circ$

REGULATION $= 1.048 - 1 \rightarrow 4.8\%$

<u>pf = 1</u> $I = 1\angle 0^\circ$ $V_2 = 1$ p.u.

$V_1 = V_2 + (1\angle 0^\circ)(.03 + j.04) = 1.031\angle 2.2^\circ$

REGULATION $= 1.031 - 1 \Rightarrow 3.1\%$

change in regulation : $4.8 - 3.1 = 1.7\%$

9. $V_{an} = 254\angle 0$, $I_a = 65.6\angle 0$

$V_{cn} = 254\angle 120^\circ$, $I_c = 65.6\angle 120^\circ$

$V_{ab} = 440\angle 30^\circ$

$V_{cb} = -V_{bc} = 440\angle 90^\circ$

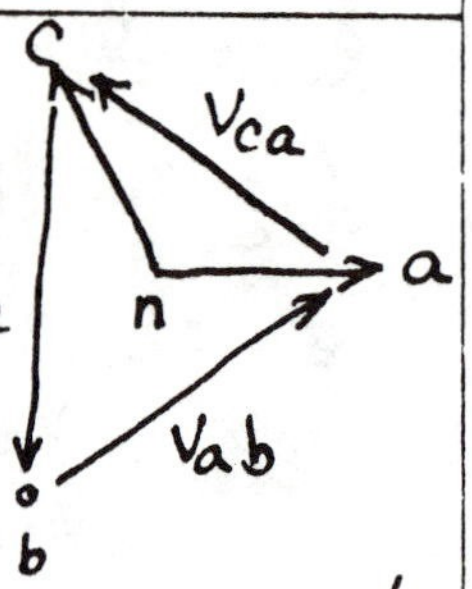

$S_{ab} = 440\angle 30^\circ \times 65.6\angle 0^\circ = 28.9\text{ KVA}\angle 30^\circ$

$S_{cb} = 440\angle 90^\circ \times 65.6\angle -120^\circ = 28.9\text{ KVA}\angle -30^\circ$

10

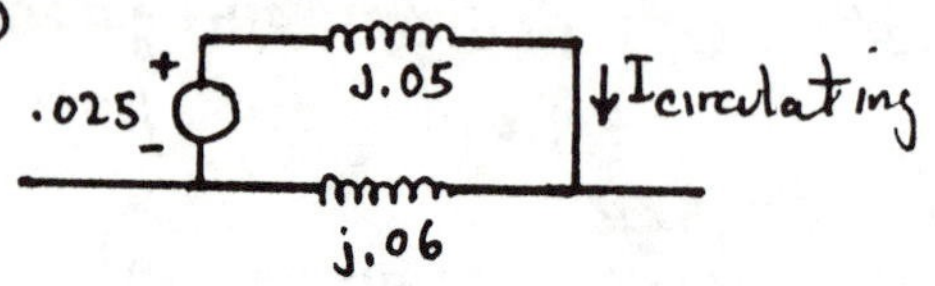

$$I_{circulating} = \frac{.025}{j.05 + j.06} = -j.227$$

CONCENTRATES

1. $$I = \frac{500\text{KW}/3\text{ phases}}{\frac{11500\text{ V}}{\sqrt{3}} \times .866} = 29\text{ A}$$

$$|S| = \frac{500\text{ KW}}{.866} = 577\text{ KVA}$$

$I_q = I\sin(\cos^{-1}.866) = 14.5\text{A}$

$$X_{comp} = -\frac{11500/\sqrt{3}}{14.5} = -458\,\Omega \text{ reactive}$$

2. First find turns ratio so that low-voltage secondary measurements can be referred to the primary:

$$a = \frac{115}{13.8} = 8.33$$

<u>Assume</u> open circuit measurements are made at rated voltage

$$I_{1\,equiv} = \frac{350}{a} = 42\text{ A}$$

$S_{oc} = V_2 I_2 = 13.8\text{KV} \times 350\text{A} = 4.83\text{ MVA}$

$$G_1 = \frac{P_{oc}}{V_1^2} = \frac{250\text{KW}}{(115\text{ KV})^2} = 1.89 \times 10^{-5}\text{ mhos}$$

$$B_1 = \frac{-\sqrt{S_{oc}^2 - P_{oc}^2}}{V_1^2} = 3.65 \times 10^{-4}\text{ mhos reactive}$$

<u>Assume</u> short circuits are at rated current: $I_{1\,rated} = \frac{50\text{ MVA}}{115\text{ KV}} = 435\text{ A}$

refer measured voltage to primary

$V_{1\,equiv} = a \times 1.5\text{KV} = 12.5\text{ KV}$

$$R_1 + a^2R_2 = \frac{P_{sc}}{I_1^2} = \frac{300\text{KW}}{435^2} = 1.585\,\Omega$$

<u>Assume</u> $R_1 = a^2R_2 = 0.79\,\Omega$

$\therefore R_2 = 0.011\,\Omega$

$S_{sc} = V_{1\,equiv} \times I_1 = 5.44\text{ mVA}$

$$X_1 + a^2X_2 = \frac{\sqrt{S_{sc}^2 - P_{sc}^2}}{I_{sc}^2} = 28.69\,\Omega$$

<u>Assume</u> $X_1 = a^2X_2 = 14.35\,\Omega$ reactive

$\therefore X_2 = .207\,\Omega$ reactive

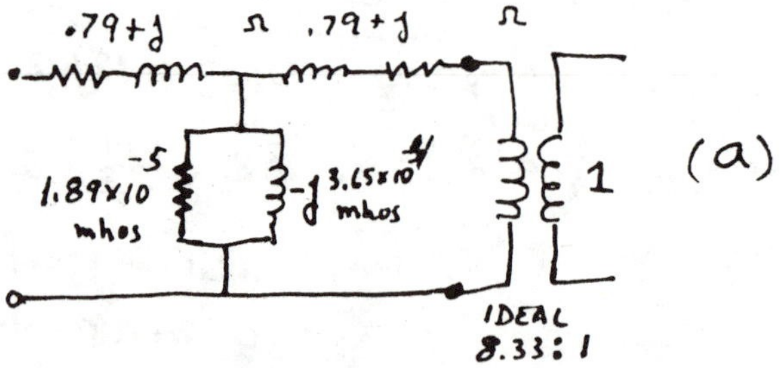

$Z_1 = a^2Z_2 = .79 + j14.35 = 14.37\angle 86.8^\circ$

$Y_c = 1.89\times 10^{-5} - j3.65\times 10^{-4} = 3.65\times 10^{-4}\angle -87^\circ$

$YZ = 5.25\times 10^{-3}\angle -.2^\circ = .005 - j0$

from fig 5.10(c)

$A = D = 1 + YZ \cong 1$

$B = Z_1 + Z_2 + YZ_1Z_2 = 28.8\angle 86.8^\circ$ (b)

$C = Y = 3.65 \times 10^{-4}\angle -87^\circ$

continued next page

2. Continued

from eq'ns 5.61 and 5.62 p 5-7

$$\frac{V_1}{I_1} = \frac{A a^2 Z_L + B}{C a^2 Z_L + D} \quad ; \quad a^2 Z_L = 0.5$$

$$Z_{in} = \frac{.5 + 28.8 \angle 86.8^\circ}{.5 \times (3.65 \angle -87) \times 10^{-4} + 1} = 28.8 \angle 85.8^\circ$$

3. from eq^{n}'s 5.65 - 5.68

$$A = (1 \times 1) + 2 Z_1 Y_C = 1 + 2(.0002 - j.001)(.01 + j.02)$$

$$= 1.00002 \angle -.00034^\circ = 1$$

$$B = 2Z_1 + 2Z_2 = .04 + j.08$$

$$C = Y_C + Y_C = .0004 - j.002$$

$$D = 2 Z_1 Y_C + 1 \rightarrow 1$$

$$a^4 Z_L = 0.1$$

$$\frac{V_1}{I_1} = Z_{in} = \frac{A a^4 Z_L + B}{C a^4 Z_L + D} = .14 + j.08$$

4. $a_{ps} = \frac{6.8\,KV}{440\,V} = 15.45$; $a_{pt} = \frac{6.8\,KV}{1.38\,KV} = 4.93$

REFERRED TO THE SECONDARY, WHERE

$$Z_{base\,Sec} = \frac{440^2}{30K} = 6.453$$

$$Z_p (P.U.) = \frac{49.5 + j110}{6.453 \times (15.45)^2} = .032 + j.071$$

$$Z_s (P.U.) = \frac{70.5 + j90}{6.453 \times (15.45)^2} = .046 + j.058$$

$$Z_t (P.U.) = \frac{50.5 + j70}{6.453 \times (15.45)^2} = .033 + j.045$$

REFERRED TO THE TERTIARY

$$Z_{base\,Ter} = \frac{1380^2}{30K} = 95.22$$

$$Z_p (P.U.) = \frac{49.5 + j110}{95.22 \times (4.93)^2} = .021 + j.048$$

$$Z_s (P.U.) = \frac{70.5 + j90}{95.22 \times (4.93)^2} = .030 + j.039$$

$$Z_t (P.U.) = \frac{50.5 + j70}{95.22 \times (4.93)^2} = .022 + j.030$$

5. Use transformer base:

$$V_{base} = \frac{440}{\sqrt{3}} \; ; \; I_{base} = \frac{250\,KVA/3}{V_{base}} = 328\,A$$

$$Z_{base} = \frac{V_{base}}{I_{base}} = 0.744\,\Omega$$

The wire resistance is:

$$500\,ft \times \frac{.053\,\Omega}{1000\,ft} = 0.0265\,\Omega$$

$$R = \frac{.0265}{.744} = .0356 \text{ p.u.}$$

$$Z_{trans} + R = .0456 + j.05 = .068 \angle 47.6^\circ$$

V_1 is transformer input. V_2 is assumed to be 1 p.u.

UNCOMPENSATED

$$I = \frac{120\,KW/3}{440/\sqrt{3}} \left(\frac{1}{328}\right) \angle -\cos^{-1}.6 = .8 \angle 53.1^\circ$$

$$V_1 = V_2 + I(Z_{trans} + R) = 1.054 \angle 0^\circ$$

COMPENSATION

rated wire current: $\frac{225}{328} = .686$

EQUATE REAL PARTS OF CURRENT

$$.8 \cos(-53.1^\circ) = .686 \times \text{new P.F.}$$

new p.f. = .7

COMPENSATED

$$I_{line} = .686 \angle -\cos^{-1}.7 = .686 \angle -45.6^\circ$$

$$V_2 = V_1 - I(R + Z_{trans})$$

$$= 1.054 \angle 0^\circ - (.686 \angle -45.6^\circ)(.37 \angle 7.8^\circ)$$

$$= 1.0074 \angle -.1^\circ$$

Then correction increases the load voltage by 0.74 %

6(a) $I_L = .855 \angle 0$, $Z = .716 \angle 65.2$

$E_g''_m = .935 - .855(j.25) = .959 \angle -12.9$

$I_m = \dfrac{E_g''_m}{j.25} = 3.84 \angle -102.9°$

$E_g''_g = .935 + .855(Z) = 1.315 \angle 25°$

$I_g = \dfrac{E_g''_g}{Z} = 1.84 \angle -40.2°$

$I_{total} = I_m + I_g = 4.95 \angle -83.7°$

BREAKER RATINGS

GEN: 1.84 X 15 MVA → 27.6 MVA @ 13.9 KV

MOTOR: 3.84 X 15 MVA → 57.5 MVA @ 13.9 KV

(b) $I_L = .855 \angle 36.9°$

$I_m = \dfrac{.935 - j.25 I_L}{j.25} = 4.308 \angle -99.1°$

$I_g = \dfrac{.935 + Z I_L}{Z} = 1.403 \angle -28.6°$

$I_{total} = I_m + I_g = 4.95 \angle -83.7°$

Breaker Ratings

GEN: 21 MVA @ 13.9 KV

MOTOR: 64.6 MVA @ 13.9 KV

(c) The total current at the fault is not affected by the power factor.

7. As in example .20, the pre-fault current is $.76 \angle 2.3°$. at point A,

$V_{th} = .9 + (.76 \angle 2.3°)(.02 + j.25)$
$= .927 \angle 11.85°$

The impedance seen from point A is:

$Z_{th} = (j.09) \| (.02 + j.25 + (j.2) \| (1.6 + j1.2))$
$= .075 \angle 89.2°$

then $I_{a1} = \dfrac{V_{th}}{Z_{th}} = 12.4 \angle -77.3$ p.u.

For a balanced fault, $I_{a0} = I_{a2} = 0$ and $I_{fault} = I_{a1}$

$|I_{fault}| = 12.4$ p.u.

8. From problem 7:

$E_{th,a} = .927 \angle 11.85°$

$Z_{th,1} = .075 \angle 89.2° = Z_1$.

$Z_2 = (j.085) \| (.02 + j.25 + (j.195) \| (1.6 + j1.2))$
$= .071 \angle 89.3°$

$Z_0 = j.6 + (j.03) \| (j.1 + (j.13) \| (1.6 + j1.2))$
$= .627 \angle 90°$

then: $I_{a1} = \dfrac{e_{th,a}}{Z_0 + Z_1 + Z_2} = 1.20 \angle -78°$

from fig. 5.17(b):

$I_{a1} = I_{a2} = I_{a0}$

$I_{fault} = I_a = I_{a1} + I_{a2} + I_{a0} = 3 I_{a1}$
$= 3.6 \angle -78°$

9. Voltages given are line-to-line

345 KV side: $I = \dfrac{750 \times 10^6/3}{345 \times 10^3/\sqrt{3}} = 1255A$

Voltage rating of common winding is

$345/\sqrt{3} = 199 KV$

500 KV side: $I = \dfrac{750 \times 10^6/3}{500 \times 10^6/\sqrt{3}} = 866A$

CURRENT RATING OF

(1) SERIES WINDING IS 866 A

(2) COMMON WINDING IS 1255 − 866 = 389 A

VOLTAGE RATING OF SERIES WINDING IS

$\dfrac{500 - 345}{\sqrt{3}} = 89.5 KV$

(3) COMMON WINDINGS ARE
.389 KA X 199 KV = 77.4 MVA

(4) SERIES WINDINGS ARE
.866 KA X 89.5 KV = 77.5 MVA

10. Choose motor base:

Assume 90% efficiency, P.F. = .85

$P_{out} = .9\, P_{in}\ ;\ VA_{in} = \frac{P_{in}}{.85}$

$VA_{in} = \frac{P_{out}}{.9 \times .85} = \frac{500 \times .746}{.9 \times .85} = 488\ KVA$

Convert system impedance to motor base

$Z_s = (.05 + j.2)\frac{488\ KVA}{100\ MVA} = .0002 + j.0010$

Convert transformer impedance to motor base

$Z_t = R(1 + j7) = .055\angle\theta$ (on own base)

$= .0078 + j.0544$

$Z_t = (.0078 + j.0544)\frac{488\ KVA}{2500\ KVA}$

$= .0015 + j.0106$

Determine motor impedance:

$I_{start} = 6 \times PF\ (P.U.) = 5.1$

$Z_m = \frac{V\ (P.U.)}{I_{start}\ (P.U.)} = j\frac{1}{5.1} = j.1961$

voltage drop (P.U.) is by the impedance ratio:

$V_{drop}(P.U.) = \frac{|Z_t + Z_s|}{|Z_t + Z_s + Z_m|} = \frac{|.0017 + j.0116|}{|.0017 + j.2077|}$

$V_{drop}(\%) = \frac{.0117}{.2077} \times 100 = 5.6\%$

TIMED

1. (a) USE TRANSFORMER BASE

$HP_{rated} = \frac{.9 \times .85 \times KVA_{rated}}{.746}$

$KVA_{rated} = 3 \times \frac{.44}{\sqrt{3}} \times I_{rated}\ (AMPS)$

max voltage drop

$.03\ (P.U.) = (6 \times .85)\, I_{rated}\ (P.U.) \times .04$

$I_{rated}(Amps) = I_{rated}(P.U.)\, I_{base}$

$I_{base} = \frac{250\ KVA/3}{.44\ KV/\sqrt{3}}$

then

$HP_{rated} = 37.7\ HP$

(b) switching from a delta to a wye connection results in 3x the line-neutral impedance, reducing the starting current by 1/3. thus a motor with 3 times the HP of (a) can be started:

$HP_{rated} = 3 \times 37.7 = 113\ HP$

(c) Voltage is reduced by .4, starting current is reduced by .4, so the motor can be increased by $1/(.4)^2$

$Hp_{rated} = \frac{37.7}{(.4)^2} = 236\ HP.$

2.(a) $V_{AB} = 440\angle 30°$

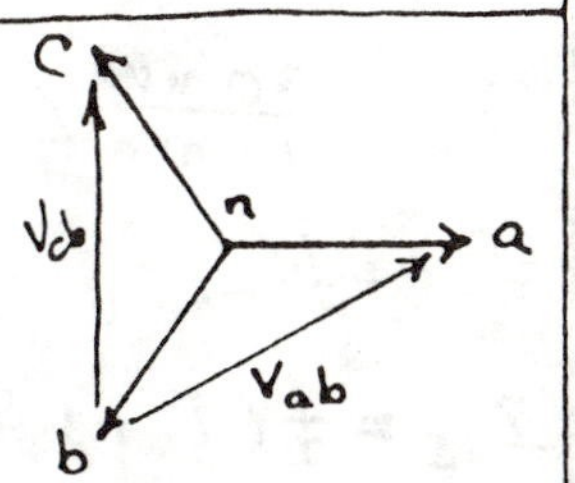

$I_a(1\phi) = \frac{10{,}000}{|V_{AB}|}\angle V_{ab}$

$= 22.73\angle 30°$

$I_a(3\phi) = \frac{50\ KVA/3}{.44\ KV/\sqrt{3}}\angle \cos^{-1} .8$

$= 65.61\angle -36.9°$

$I_b(3\phi) = 65.61\angle -156.9°$

$I_c(3\phi) = 65.61\angle 83.1°$

$I_a = I_a(1\phi) + I_a(3\phi) = 77.4\angle -21.2°$

$S_{ab} = V_{ab} \times I_a^* = 34.1\angle 51.2°$

$S_{bc} = V_{cb} \times I_c(3\phi)^* = 28.9\ K\angle 6.9°$

2. continued

(b) $I_c = 65.6\angle 83.1^\circ + 22.73\angle 90^\circ$

$= 88.21\angle 84.9^\circ$

$S_{cb} = (440\angle 90^\circ)(88.21\angle -84.9^\circ) = 38.8\angle 5.1^\circ$

$S_{ab} = (440\angle 30^\circ)(65.6\angle 36.9^\circ) = 28.9K\angle 66.9^\circ$

3. USE 25MVA BASE

$Z_{15} = j.06 \times \frac{25MVA}{15MVA} = j.1$

(a) with equal reactances, both carry the same current.

(b)

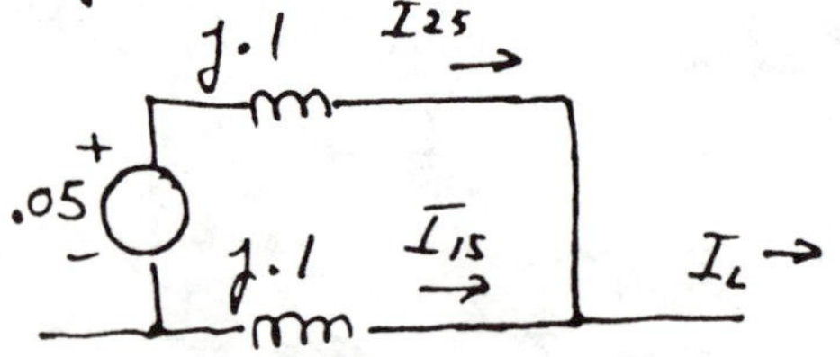

KCL: $I_{15} + I_{25} = I_L$

KVL: $.05 = j.1 I_{25} - j.1 I_{15}$

$I_{25} - I_{15} = -j.5$

$I_L = \frac{30\,MVA}{25\,MVA}\angle -\cos^{-1}.8 = 1.2\angle -36.9$

$I_{25} = \frac{1}{2}(I_L - j.5) = .776\angle -51.8^\circ$ p.u.

$I_{15} = \frac{1}{2}(I_L + j.5) = .492\angle -12.9^\circ$ p.u.

KVL: $.05 = (j.1 + j.1) I_{circ}$

$I_{circ} = -j.25$ p.u.

CURRENT BASE: $\frac{25 \times 10^6}{\sqrt{3} \times 35 \times 10^3} = 412\,A$

$I_{25} = 412 \times .776\angle -51.8 = 320\angle -51.8^\circ A$

$I_{15} = 412 \times .492\angle -12.9^\circ = 203\angle -12.9^\circ A$

$I_{circ} = 412 \times j(-.25) = 103\angle -90^\circ A$

4. from Concentrate #8

$e_{th,a} = .927\angle 11.85^\circ$

$Z_1 = .075\angle 89.2$

$Z_2 = .071\angle 89.3$

from fig 5.18

$-I_{a2} = I_{a1} = \frac{e_{th,a}}{Z_1 + Z_2} = 6.35\angle -77.4^\circ$ p.u.

eq 5.150 $\quad I_{fault} = j\sqrt{3} I_{a1} = 11.0\angle 12.7^\circ$

eq 5.144 with $V_{a0} = 0$

$V_a = -2V_b$

eq 5.145 $\quad 3V_{a1} = V_a + (a + a^2)V_b = -3V_b$

fig 5.18 $\quad V_{a1} = \frac{Z_2}{Z_1 + Z_2} e_{th,a} = .464\angle 11.85^\circ$

$V_b = -.464\angle 11.9^\circ$

$V_a = .927\angle 11.9^\circ$

5. As no current flows when bus 3 is open, $e_{th,a} = 1\angle 0$

To find thevenin impedance

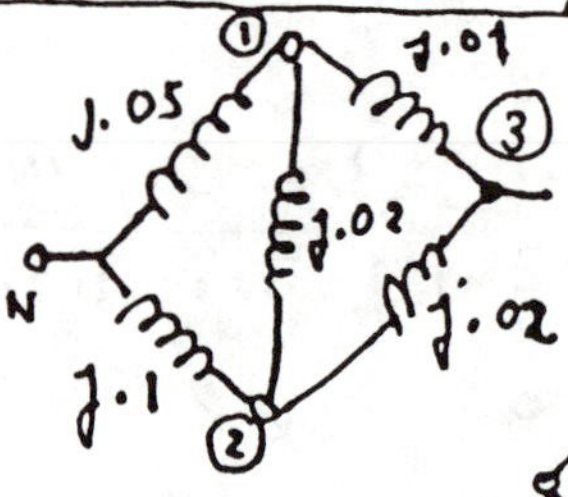

→ Use Δ-Y transform to obtain

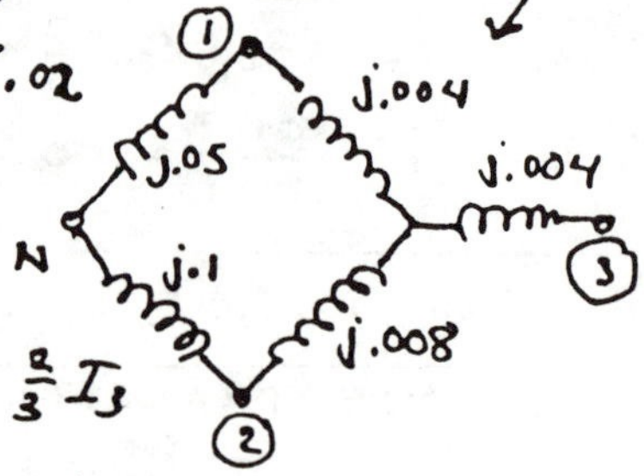

note that

$I_1 = \frac{.108}{.108 + .054} I_3 = \frac{2}{3} I_3$

$I_2 = \frac{.054}{.108 + .054} I_3 = \frac{1}{3} I_3$

$Z_{th} = Z_1 = Z_2 = j.004 + (j.108 \| j.054)$

$= j.04$

fig. 5.18

$-I_{a2} = I_{a1} = \frac{e_{th}}{Z_1 + Z_2} = 12.5\angle -90^\circ$

continue

5. continued

from eq. 5.150 $I_{fault} = j\sqrt{3} I_{a_1} = 21.65 \angle 0°$

from the note above, the fault current divides such that the current flowing from the left source into bus #1 is 2/3 x I_{fault}, and from the right source into bus #2 is 1/3 x I_{fault}.

At Bus #1: $V_{a1} = 1 \angle 0$ (no current)

$$V_{1b} = 1 \angle{-120°} - j.05 \times \tfrac{2}{3} I_{fault}$$

$$= 1.67 \angle{-107.5°} \text{ p.u.}$$

$$V_{1c} = 1 \angle 120 + j.05 \times \tfrac{2}{3} I_{fault}$$

$$= 1.67 \angle 107.5° \text{ p.u.}$$

At Bus #2; $V_{2a} = 1 \angle 0$

$$V_{2b} = 1 \angle{-120°} - j.1 \times \tfrac{1}{3} I_{fault}$$

$$= 1.67 \angle{-107.5°}$$

$$V_{2c} = 1 \angle 120° - j.05 \times \tfrac{1}{3} I_{fault}$$

$$= .711 \angle 134.7°$$

6. The transformer primary is rated at 60 MVA (H-X) + 21 MVA (H-T) = 81 MVA. It is assumed that all impedances are on the primary base:

$$Z_{base} = \frac{(138 \text{ KV})^2}{81 \text{ MVA}} = 235 \Omega$$

$$Z_\ell = \frac{j\,3.1}{Z_{base}} = j\,.013$$

(a) for a 3-phase fault at the X-winding, T is not involved, and is assumed to be an open circuit. The impedance involved is then $Z_\ell + Z_{H\text{-}X} = j(.013 + .098) = j.111$

so $I_{fault} = \dfrac{1 \text{ p.u.}}{j.111} = -j\,9.02$ p.u.

The current base at t X-winding is

$$\frac{(81,000/3) \text{ KVA}}{34.5 \text{ KVA}/\sqrt{3}} = 1356 \text{ A}$$

so $|I_x| = 1356 \times 9.02 = 12,230$ A

(b) For a single-phase fault to ground (see fig. 515(e)) with both X and H in a grounded neutral configuration with one phase to ground,

$Z_0 = Z_1 = Z_2 = j.111$; then from fig. 517(b)

$$I_{a1} = \frac{e_{th,a}}{Z_1 + Z_2 + Z_3} = \frac{1}{3 \times j.111} = -j3$$

eq. 5.136: $I_a = 3 I_{a_1} = -j\,9$ p.u.

then $I_{xa} = 9 \times 1356 = 12,204$ A

7.

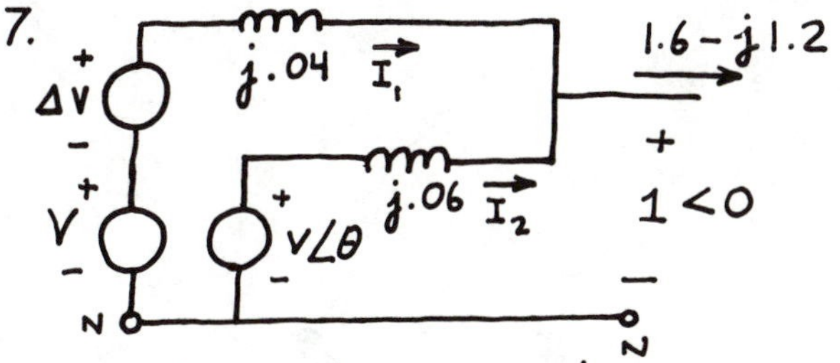

A transformer base is used.

$$I_1 = \frac{V(1+\Delta) - 1}{j.04} \;;\; I_2 = \frac{V(\cos\theta + j\sin\theta) - 1}{j.06}$$

$$I_1 + I_2 = 1.6 - j\,1.2$$

As $|I_1| \le 1$ and $|I_2| \le 1$ and $|I_1 + I_2| = 2$ then it is necessary that $|I_1| = 1$, $|I_2| = 1$ and $I_1 = I_2 = .8\,j.6$

then, from Kirchhoff's voltage law:

$$V(1+\Delta) = j.04(.8j.6) + 1 = 1.024 + j.032 = 1.0245 \angle 1.790° \quad ①$$

$$V \angle \theta = j.06(.8j.6) + 1 = 1.036 + j.048 = 1.0371 \angle 2.653° \quad ②$$

obtaining the magnitude from ② and the angle from ①

$$V = 1.0371 \angle 1.790°$$

from ② $\theta = .863°$

from ① $1 + \Delta = 1.0245/1.0371$ or

$$\Delta = -.01215$$

as $\Delta = .1n/16$ $\quad n = -1.94 \rightarrow -2$

8. As the load is $\frac{34\text{ MW}}{.85\text{ p.f.}} = 40\text{ MVA}$, the transformer must be on the 50MVA setting. Then $Z_{base}(\text{tr}) = \frac{(12\text{ KV})^2}{50\text{ K}^2\text{VA}} = 2.88\,\Omega$

For the transmission line

$$Z_{base}(\text{line}) = \frac{(12.47\text{ KV})^2}{100\text{ K}^2\text{ VA}} = 1.555\ \Omega$$

Using the transformer secondary base:

$$Z(\text{line}) = \frac{1.555}{2.88}(.02 + j.1) = .011 + j.054 \text{ p.u.}$$

$$Z(\text{tr}) = R(1 + j15) \qquad R = \frac{.075}{|1 + j15|} = .005$$

$$Z(\text{tr}) = .005 + j.075 \text{ p.u.}$$

$$V_{LOAD} = \frac{12.16}{12} = 1.013 \text{ p.u.}$$

$$I_{LOAD} = \frac{40\text{ MVA}/50\text{ MVA}}{1.013}\angle -\cos^{-1}.85 = .789\angle -31.8^\circ$$

$$Z = Z(\text{tr}) + Z(\text{line}) = .016 + j.129 = .13\angle 82.9^\circ$$

$$V_s = IZ + 1.013 = 1.08\angle 4.2^\circ$$

Regulation $= 1.070 - 1.013 = .057$ p.u

$= 5.7\%$

9.

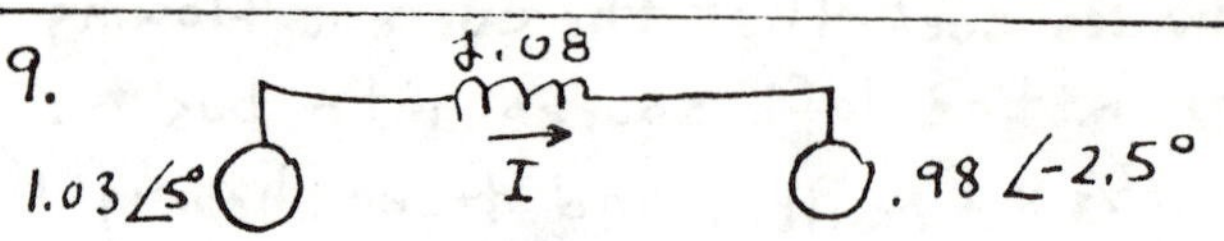

$$I = \frac{1.03\angle 5^\circ - .98\angle -2.5^\circ}{j.08} = 1.758\angle -19.5^\circ$$

$$P + jQ = VI^* = 1.03\angle 5^\circ \times 1.758\angle +19.5^\circ$$

$$P + jQ = 1.65 + j.75 \text{ P.U.}$$

10.

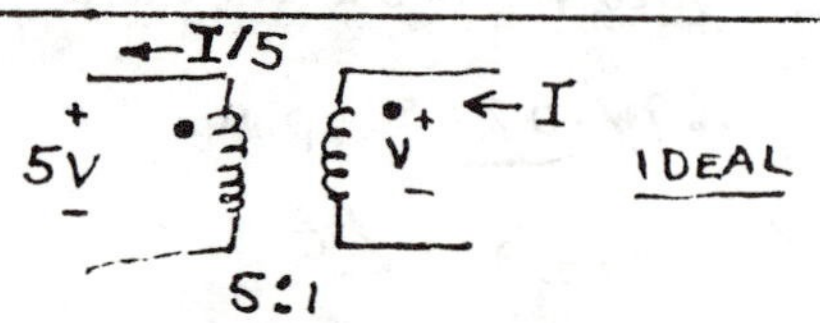

Correct

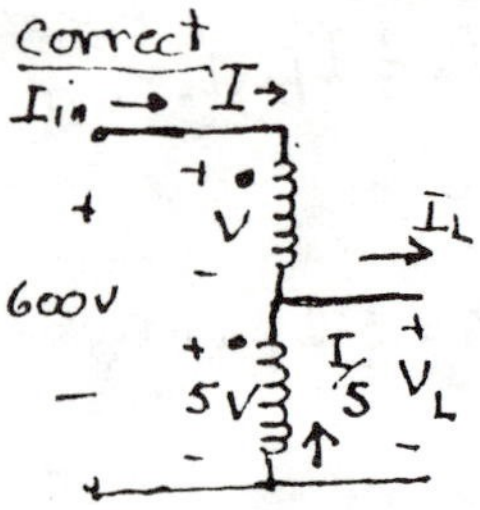

$I_L = I + I/5 = \frac{6}{5} I_{in}$

$V_L = 5V = \frac{5}{6} \times 600 = 500$

100 V winding: I_{rated}

500 V winding: $.2\, I_{rated}$

output: $1.2\, I_{rated}$

rated LOAD: $|Z_L| = \frac{500}{1.2\, I_{rated}} = \frac{417}{I_{rated}}$

Incorrect

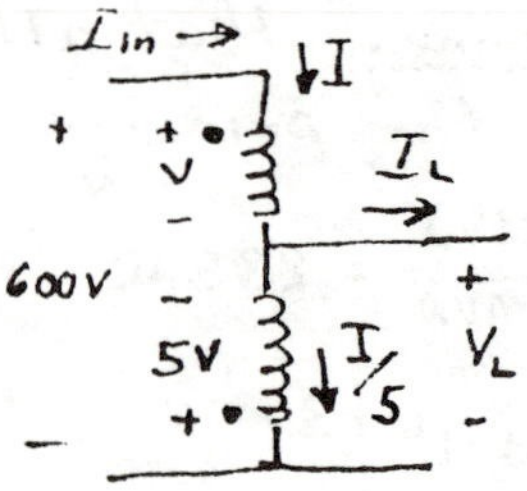

$-5V = V_L$

$600 = V - 5V = -4V$

$= \frac{4}{5} V_L \quad \therefore V_L = 750$

$I_L = I - \frac{I}{5} = \frac{4}{5} I$

$I_{in} = \frac{5}{4} I_L$

for rated impedance

$$|I_L| = \frac{750}{|Z_L|} = \frac{750 \times 1.2\, I_{rated}}{500} = 1.8\, I_{rated}$$

$$|I_{in}| = \frac{5}{4}|I_L| = 2.25\, I_{rated}$$

500 v winding: $|I| = \frac{2.25}{5} I_{rated}$

$= 2.25\,(.2\, I_{rated})$

both windings have 125% over current

WARM UPS

1. $\delta = \sqrt{\dfrac{1.8\times10^{-8}}{\pi\times10^{7}\times(4\pi\times10^{-7})}} = 2.135\times10^{-5}\ m$

$r = \dfrac{.2}{2} = .10\ cm = 10^{-3} m\ ;\ R = 1 cm = 10^{-2} m$

$Z_i = (1+j)\left[\dfrac{\rho}{\delta\, 2\pi r} + \dfrac{\rho}{\delta\, 2\pi R}\right] = \dfrac{\rho(1+j)}{2\pi\,\delta}\left[\dfrac{1}{r}+\dfrac{1}{R}\right]$

$= \dfrac{1.8\times10^{-8}(1+j)}{2\pi\times2.135\times10^{-5}} = 0.1476(1+j)$

2. AWG #1 D = 289 mils (Appendix A)

$r = 144.5$

at 10.0 KHz for copper: δ = 26 mils (ex. 6.1)

then $\dfrac{r}{\delta} = 5.56$: eq 6.8 $\dfrac{R}{R_o} = .25 + .5\dfrac{r}{\delta} = 3.03$

eq 6.9 $\dfrac{X_i}{R_o} = .5025\dfrac{r}{\delta} = 2.79$

from ex. 6.3 $R_o = .026\ \Omega/1000$ ft

$Z_i = .026(3.03 + j2.79) = .079 + j.073$

3. $r = 16$ mils ; $\delta = 2.6$ mils (ex. 6.2)

$\dfrac{r}{\delta} = 6.15$: eq 6.9 $\dfrac{X_i}{R_o} = .5025\dfrac{r}{\delta} = 3.09$

$R_o = 10.4\ \dfrac{\Omega}{1000\ ft}$ (Appendix A)

X_i per conductor $= 3.09\ R_o = 32.16\ \Omega$

X_i total $= 64.32\ \Omega$

$\dfrac{D}{r} = \dfrac{1\ cm}{16\ mils}\times\dfrac{10^3\ mils}{inch}\times\dfrac{1\ inch}{2.54\ cm} = 24.606$

$L_e = 4\times10^{-7}\ln\dfrac{D}{r} = 12.81\times10^{-7}\ Hy/m$

$X_e = 2\pi f L_e = 2\pi\times10^6 L_e = 8.05\ \Omega/m$

$= 8.05\dfrac{\Omega}{m}\times\dfrac{1m}{100cm}\times\dfrac{2.54cm}{in}\times\dfrac{12000''}{1000ft} = 2454\ \Omega$

then

$X_{total} = X_e + X_i = 2518\ \Omega/1000$ ft

4. $C = \dfrac{10^{-9}}{18}\ln\dfrac{2}{.2} = 2.41\times10^{-11} F/m = 24.1\ pF/m$

5. TABLE 6.2 : GMR = .0586 ; X_L(1 ft) = .344 } bluebird

$K_L = 1 + \dfrac{\ln D}{\ln(1/GMR)} = 1 + \dfrac{\ln 10}{\ln\frac{1}{.0586}} = 1.812$

$X_L = 1.812 \times .344 = .623\ \Omega$/mile

6. $K_C = 1 + \dfrac{\ln(20\ ft)}{\ln\left(\frac{12}{1.762/2}\right)} = 2.147$

X_C × miles $= .0776 \times K_C = .167$ MΩ-miles/cond

for 2 miles

X_C/conductor $= \dfrac{.167}{2} = .083$ MΩ/cond

for 2 conductors

$X_C = \dfrac{.083}{2} = .0415$ MΩ

7. As O.D. is in inches, spacing of 1 ft

$\ln\dfrac{d}{r} = \ln\dfrac{d(ft)}{\frac{O.D.(in)}{2}\times\frac{1}{12}} = \ln\dfrac{24\,d(ft)}{O.D.(in)}$

$X_C = 29688\ \ln\dfrac{24\times1}{O.D.(in)}$

	CALCULATED X_C	TABLE X_C
WAXWING	.1091 MΩ-mi	.1090 MΩ-mi
Partridge	.1075 " "	.1074 " "
bluebird	.0776 " "	.0776 " "

etc.

8. P 4-5

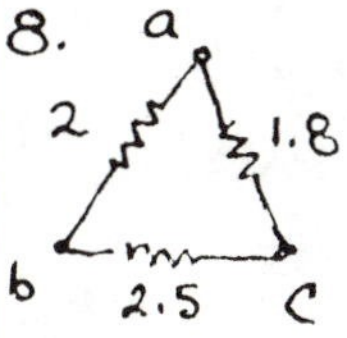

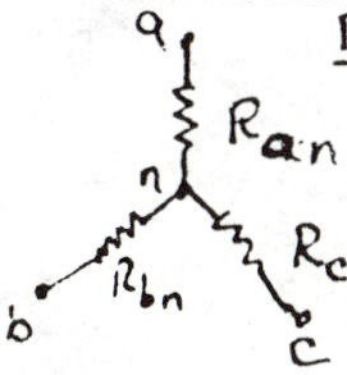

$R_{an} = \dfrac{2\times1.8}{2+1.8+2.5} = .571$

$R_{cn} = \dfrac{2.5\times1.8}{6.3} = .714$

$R_{bn} = \dfrac{2\times2.5}{6.3} = .794$

9. The Volt-Ampere Reactive rating is the same for both. The rated

Voltages: $V_\Delta = \sqrt{3}\ V_Y$

rated currents $I_\Delta = \dfrac{1}{\sqrt{3}} I_Y$

$C \propto \dfrac{I}{V}$: $\dfrac{I_\Delta}{V_\Delta} = \dfrac{1}{3}\dfrac{I_Y}{V_Y}$

Δ takes higher voltage rating, lower capac

CONCENTRATES

1. from table 6.2 : GMR = .0217

$d_s = \sqrt[3]{10 \times 10 \times 10\sqrt{2}} = 11.22$ ft.

using eq 6.34 and example 6.8

$2\pi f L = 120\pi \times 2\times 10^{-7} \ln \frac{d_s}{GMR} = 4.71 \times 10^{-4}\ \Omega/m$

$X_{ab} = X_{ca} = 120\pi \times 2 \times 10^{-7} \ln \frac{10\times 10\sqrt{2}}{d_s^2} = 8.77\times 10^{-6}\ \Omega/m$

$X_{bc} = 120\pi \times 2 \times 10^{-7} \ln \frac{10 \times 10}{d_s^2} = -1.74 \times 10^{-5}\ \Omega/m$

as there are 1609 meters/mile

$X_L = 1609 \times 2\pi f L = .758\ \Omega/mi$

$X_{ab} = X_{ca} = 1609 \times 8.77 \times 10^{-6} = .014\ \Omega/mi$

$X_{bc} = 1609 \times 1.74 \times 10^{-5} = -.028\ \Omega/mi$

2. Transform line-to-line impedance to line-neutral equivalent:

$Z = Z_{line} + Z_{load} = (3 + j4) + (\frac{20}{3} + j5)$

(a)

$I^2 \frac{20}{3} = \frac{30{,}000}{3} \qquad I^2 = 1500$

use I as reference: $I = \sqrt{1500}\angle 0°$

$V_s = IZ = \sqrt{1500}\,(\frac{29}{3} + j9) = 512\angle 43°$

$S_s = V_s I^* = 1500(\frac{29}{3} + j9)$

P = 14.5 KW Q = 13.5 KVAR

(b)

$I = \sqrt{1500}\angle 0°$

$V_s = \sqrt{1500}\,(\frac{29}{3} + j4) = 405\angle 22.5°$

$S_s = P + jQ = 1500(\frac{29}{3} + j4)$

P = 14.5 KW Q = 6 KVAR

note: these are per-phase values

$P_{total} = 43.5$ KW, Q = 18 KVAR

3. As in example 6.9:

$Z_{line} = 4.86 + j17.2 = 17.87\angle 74.2°$

Take the receiving end voltage as reference: $V_R = V\angle 0°$

then

$11\text{KV}\angle\theta = V_R + (.225\angle \cos^{-1}.9)\, Z_{line}$ (KV)

$= V + (-.701 + j3.959)$

$11\cos\theta + 11j\sin\theta = V - .701 + j3.959$

EQUATE IMAGINARY PARTS:

$11\sin\theta = 3.959$; $\sin\theta = .36$; $\theta = 21.1°$

$V = 11\cos\theta + .701 = 10.963$ KV

$\%\,Reg = \frac{11 - 10.963}{10.963} \times 100 = .34\%$

4.

	O.D. (in.)	R/mi 60 Hz 50°C	GMR (ft.)	X_L Ω/mi	1-ft spacing X_C MΩ-mi
Falcon	1.545	.0667	.0523	.385	.0814

$K_L = 1 + \frac{\ln 20}{\ln \frac{1}{GMR}} = 2.015$; $K_C = 1 + \frac{\ln(20(12))}{\ln \frac{24}{OD(in)}} = 2.092$

$X_L = K_L \times .358 = .7214\ \frac{\Omega}{mi}$; $X_C = K_C \times .0814 = .1703 \times 10^6\ \Omega\text{-}mi$

$Z = .0667 + j.7214\ \frac{\Omega}{mi}$; $Y = j\frac{10^{-6}}{.1703} = 5.8724 \times 10^{-6}\ ℧/mi$

(a) $YZ = 4.2544 \times 10^{-6}\angle 174.7°$

$\sqrt{YZ} = 2.0626 \times 10^{-3}\angle 87.4° = \gamma$

(b)

$\gamma = \alpha + j\beta = 9.505 \times 10^{-5} + j2.0604 \times 10^{-3}$

$\alpha = 9.505 \times 10^{-5}$

(c) $\beta = 2.0604 \times 10^{-3}$

(d) $\frac{Z}{Y} = 1.2338 \times 10^{5}\angle -5.28°$

$Z_0 = \sqrt{\frac{Z}{Y}} = 351.25\angle -2.6° = 350.88 - j16.19$

(e) $\rho = \frac{Z_L - Z_0}{Z_L + Z_0} = \frac{100 - 350.88 + j16.19}{100 + 350.88 - j16.19}$

$= .557\angle 178.4° = -.557\angle -1.6°$

5. $\rho = \frac{70-50}{70+50} = \frac{1}{6}$ (a)

(b) $Z_L = 50 \frac{70\cos\pi/2 + j50\sin\pi/2}{50\cos\pi/2 + j70\sin\pi/2} = \frac{2500j}{70j} = 35.714\,\Omega$

(c) $VSWR = \frac{1+|\frac{1}{6}|}{1-|\frac{1}{6}|} = \frac{7}{5} = 1.4$

6. $z_\ell = \frac{25-j25}{50} = .5 - j.5$

On a ρ = constant curve, centered on $r=1$, $x=0$, the intersection with $r=1$ circle is at $z = 1+j1$. On a Smith chart this is 180° toward the source from the load point. The compensation impedance is then $-j1$ (a capacitance)

$\frac{180}{720}\lambda = \frac{\lambda}{4}$ toward the source.

$Z_{comp} = -j1 \times z_0 = -j50\,\Omega$

7.(a) $Z_{in} = -j25 = z z_0 \therefore z = -j.5$

An open circuit ($Z_L = \infty$) is at the bottom of the Smith chart (Appendix C). Traversing clockwise (toward the generator), $z = -j.5$ is 126.5° from the open. This is $\frac{126.5}{720}\lambda = .176$ wavelengths.

(b) $Z_{in} = j50 = z z_0 \therefore z = j1$.

A short circuit ($z = 0$) is at the top of the Smith Chart. Traverse 90° to reach the $z = j1$ point. Thus the cable length is $\frac{90}{720}\lambda$ or 1/8 wave length

TIMED

1. $V_{\ell-n} = \frac{86.6\text{ KV}}{\sqrt{3}} = 50\text{ KV}$ at receiving END

$|I_R| = \frac{8500/3\text{ KVA}}{50\text{ KV}} = 56.67\text{A}$

$I_R = 56.67\angle -\cos^{-1}\frac{75}{85} = 56.67\angle -28.1°$

As in example 6.10

$Y = j6.37\times10^{-4}$ mhos ; $Z = 21.34 + j68.8\,\Omega = 72.034\angle 72.8°$

so

$1 + YZ = .9781 + j.0068 \approx .9781$

$V_S = .9781 \times 50\text{KV} + (56.67\angle -28.1°)(72.034\angle 72.8°)\times10^{-3}$

$= 51.81\angle 2.9°$

$|V_R (N.L.)| = \frac{51.81}{.9781} = 52.97$

% Regulation $= \frac{52.97-50}{50}\times100 = 5.94\%$

2. $z_0 = 72$, $Z_L = 50$, $\beta = .5$, $\ell = 7.5$

$z_L = \frac{50}{72} = .69$ $\beta\ell = 3.75\text{ rad} = 215°$

$2\beta\ell = 430 = 360 + 70$

the load point on the Smith chart is at $.69 + j0$ above the center of the circle. the source point is 70° clockwise on the constant $|\rho|$ circle. At this point:

$z_{in} = .82 + j.31$

$Z_{in} = 72 z_{in} = 59 + j22.3$

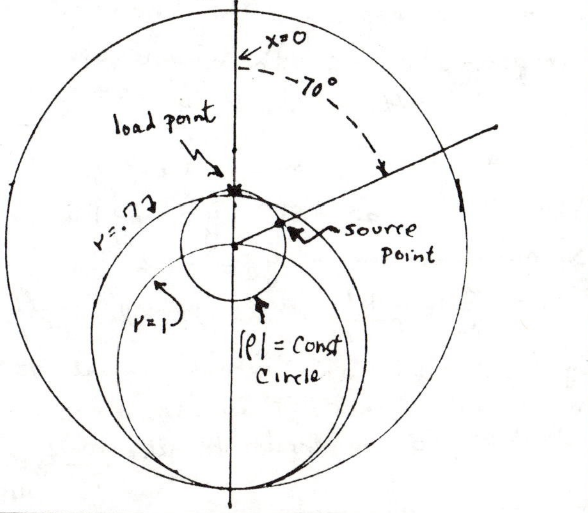

3. $z_o = 100$, $Z_L = 25 + j25$

$\therefore z_L = \frac{Z_L}{z_o} = .25 + j.25$; $y_L = \frac{z_o}{Z_L} = 2 - j2$

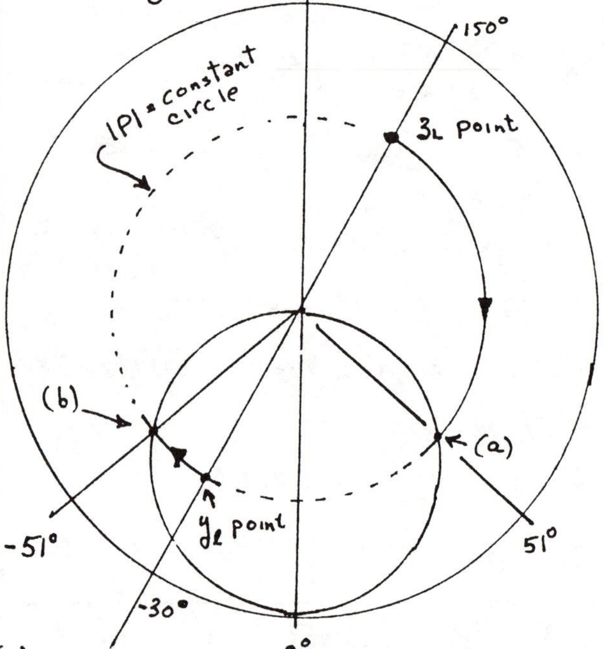

(a)

z_L point $.25 + j.25$ is at 150°

intersection for compensation is at $z = 1 + j1.6$ at 51° ((a) above)

location of series compensation point at $\frac{150-51}{720}\lambda = .0.1375\lambda$.

The required compensation is

$Z_c = -j1.6 \times 100 = -j160\,\Omega$.

for the 72Ω line $z_c = -j\frac{160}{72} = -j2.22$

On smith chart this is at −48° on the periphery. the shortest Compensator is then an open (at 0°) with a length of $\frac{48}{720}\lambda = \frac{\lambda}{15}$.

(b) y_L at −30°, compensation point at (b) on sketch, at −51° $y = 1 - j1.6$

so compensation is $y_c = j1.6$ or

$Y_c = \frac{y}{Z_o} = j\frac{1.6}{100}$; for 72Ω line $y_c = \frac{j1.6}{100} \times 72$

$y_c(72\Omega) = j1.152$, which is at 73° on smith Chart. This uses an open (Y=0) at 180°, so compensator length $= \frac{180-73}{720}\lambda = .1486\lambda$

4. $Z_L = 25 - j25$ $z_o = 50$

$y_L = \frac{z_o}{Z_L} = 1 + j1$ at 63° on Smith chart

follow constant |ρ| circle around to $y = 1 - j1$ at −63° on Smith chart

This is $\frac{63-(-63)}{720}\lambda = .175\lambda$ from load.

Compensating susceptance is $b_c = +j1$

or $B_c = \omega C = \frac{b_c}{z_o} = \frac{1}{50} = .02$

then $C = \frac{.02}{\omega}$

5. As parameters are given as PI Parameters, assume fig 5.10(d) with

$Y_1 = Y_2 = Y_{shunt} = Y$ $A = 1 + YZ$

$Y = .001063j$ $B = Z$

$Y(p.u.) = Y Z_{base}$ * $Z_{base} = \frac{(220KV)^2}{50\,MVA/.9}$ *

* LOAD BASE

$= 871.2\,\Omega$

$Y(p.u.) = j.9261$ $Z(p.u.) = \frac{40 + j160}{Z_{base}}$

$= .1893 \angle 76°$

$A = 1 + YZ = .8310 \angle 3°$

$V_s = AV_R + BI_R$ $V_R = 1\angle 0°$, $I_R = 1\angle -\cos^{-1}.9$

$V_s = (.831\angle 3°)1 + (.1893\angle 76)(1\angle -\cos^{-1}.9)$

$= .9698 \angle 11.2°$ p.u.

$|V_s| = .9698 \times 220KV = 213.4\,KV$

6. $X_C = -.15 \times 10^6\,\Omega\text{-mi} \times \frac{1}{100\,mi} = -.15 \times 10^4\,\Omega$

$B_C = 6.667 \times 10^{-4}$; $Z_{base} = \frac{(132KV)^2}{75.mw/.92} = 213.7\,\Omega$

$Y = jB_C \times Z_{base} = 0.142j$ (p.u.)

$Z = \frac{100(.1 + j.6)}{213.7} = .2846 \angle 80.5°$

$A = 1 + \frac{YZ}{2} = .980 \angle 0°$ $B = Z$

$V_s = AV_R + BI_R$; $V_R = 1\angle 0$, $I_R = 1\angle -\cos^{-1}.92$

$= (.980\angle 0°)(1\angle 0°) + (.2846\angle 80.5°)(1\angle -\cos^{-1}.92)$

$= 1.158 \angle 11.9°$ p.u.

$V_s = 132KV \times 1.158\angle 11.9°$

$= 153KV \angle 11.9°$

$|V_s| = 153\,KV$

WARM UPS

1. $R_A = \frac{V_{n.L.} - V_{f.L.}}{I_{f.L.}} = \frac{260-250}{50} = 0.2\,\Omega$

2. $I_{f.L.} = \frac{10\,KW}{240\,V} = 41.67\,A$; $R_A = \frac{260-240}{41.67} = .48\,\Omega$

$P_{in} = P_{out} + I^2 R_A + P_{mech\,loss}$

$P_{mech\,loss} = 15\,HP \times \frac{746\,W}{HP} - 10\,KW - .48(41.67)^2$

$= 357$ Watts

3. $T = k\phi I_A$, $E_g = k\phi\Omega$, $\phi = \sim V_s$

$V_s = I_A R_A + E_g$ $P_{in} = V_s I_A$ $P_{out} = T\Omega$

(a) T = constant

$V_s \downarrow 10\%$, $\phi \downarrow 10\%$ ∴ $I \uparrow \frac{1}{.9}$

$P_{in} = V_s I_A$ = constant

$P_{out} = T\Omega = P_{in} - I_A^2 R_A - P_{mech}$

a P_{in} is constant & $P_{mech} \approx$ const

then P_{out} drops by the increase in $I_a^2 R_a$, which is on the order of 1%, so Ω drops about 1%

(b) $T \propto \Omega$

$V_s = I_a R_a + E_g = \frac{R_a T}{k\phi} + k\phi\Omega$

let $\frac{R_a T}{k} = b\Omega$, $\phi = aV_s$

$V_s = \left(\frac{b}{\phi} + k\phi\right)\Omega$

$V_s = \left(\frac{b}{aV_s} + aV_s k\right)\Omega$

$\Omega = \frac{1}{ak + \frac{b}{aV_s^2}}$

when $V_s \downarrow \times .9$ $\Omega \downarrow \sim .995$

$I = \frac{T}{k\phi} \propto \frac{\Omega}{V}$, $I \uparrow \sim \frac{.995}{.9} \sim 10.5\%$

$P_{in2} = P_{in1} \times 1.105 \times .9$, drops $\sim .5\%$

4. $E_g = K\phi\Omega$ $T = K\phi I_A$

as $\phi \propto I_a$ let $K\phi = K' I_A$

then

$E_g = K' I_A \Omega$, $T = K' I_A^2$

if T increases by 1.25, I_A increases by $\sqrt{1.25} = 1.118$ or 11.8%

As $E_g \approx V_s$ = constant, then

$E_g \approx$ constant $= K' I_A \Omega$

so $I_{A2}\Omega_2 = I_{A1}\Omega_1$, or $\frac{\Omega_2}{\Omega_1} = \frac{I_{a1}}{I_{a2}}$

or $\frac{\Omega_2}{\Omega_1} = \frac{1}{1.118} = .894$ or Ω drops 10.6%

P_{in} increases as I_A, 11.8%

$P_{out} = T\Omega \to T_1\Omega_1 \times 1.25 \times .894 \sim 11.8\%$

5. $P/_\phi = V_{an}^2 g_c +$ rot losses $+ I_r^2 (r_s + \frac{a^2 r_r}{s})$

$I_r^2 (r_s + \frac{a^2 r_r}{s}) = \frac{25{,}000 - 850}{3}$

$I_r^2 r_s = \frac{550}{3}$ ∴ $I_r^2 \frac{a^2 r_r}{s} = 7867\,W$

$\Omega_s \to 1800$ rpm ∴ $s = \frac{1800-1700}{1800} = \frac{1}{18}$

then $I_r^2 a^2 r_r = 437$

$P_L/_\phi = I_r^2 a^2 r_r \left(\frac{1}{s} - 1\right) = 7429$ W/phase

$P_L = 22{,}287\,w = 29.9$ H.P.

Efficiency: $\eta = \frac{22287}{25000} \times 100 = 89.1\%$

6. $r_c = \frac{1}{g_c} = \phi \frac{V_{L-n}^2}{P_{n.L.}} = 3\frac{(440/\sqrt{3})^2}{1200} = 161\,\Omega$

$x_\phi = \frac{-1}{b_c} = \phi \frac{V_{L-n}^2}{(S_{nL}^2 - P_{nL}^2)^{1/2}} = \frac{3 \times (\frac{440}{\sqrt{3}})^2}{(2500^2 - 1200^2)^{1/2}} = 88.3\,\Omega$

7. $(21)^2 R_e = \frac{375}{3}$ ∴ $R_e = .2834\,\Omega$

$(21)^2 X_e = \frac{Q}{3}$ $Q = \frac{P}{pf}\sqrt{1 - pf^2} = 941.6$

$X_e = .7117\,\Omega$

8. $\frac{P_L}{q} = \frac{20\,HP \times 746\,\frac{W}{HP}}{3\,phases} = 4973\,\frac{W}{phase}$

$= \frac{I_r'^2 a^2 r_r (1-S)}{S} = I_r'^2 a^2 r_r \left[\frac{1}{.03} - 1\right]$

$I_r'^2 a^2 r_r = \frac{.03}{.97} 4973 = 153.8$

$\frac{P_{in} - P_{fixed}}{3} = \frac{16950 - 1200}{3} = 5250 = I_r'^2 \left(r_s + \frac{a^2 r_r}{s}\right)$

$\frac{5250}{I_r'^2 a^2 r_r} = \frac{5250}{153.8} = \left(\frac{r_s}{a^2 r_r} + \frac{1}{s}\right) = 34.135$

or $\frac{r_s}{a^2 r_r} = 34.135 - \frac{1}{.03} = .8019$

from 7: $R_e = .2834 = r_s + a^2 r_r$

$= a^2 r_r \left(\frac{r_s}{a^2 r_r} + 1\right)$

$\therefore \quad a^2 r_r = .157\,\Omega$

$r_s = .126\,\Omega$

9. $V_A I_A = j\frac{Q}{3} = \frac{V_A (V_A - E_g)}{j X_s} \therefore E_g = V_A + \frac{Q X_s}{3 V_A}$

$E_g = 254 + \frac{10{,}000 \times 20}{3 \times 254} = 516$ Volts

$I_r = \frac{5A}{254V} \times 516V = 10.16A$

10. $V_A = \frac{13800}{\sqrt{3}} = 7967$; $|I_A| = \frac{50 \times 10^6/3}{1.38 \times 10^4/\sqrt{3}} = 2092$

$\angle I_A = -\cos^{-1}.88 = -28.4°$

$jX_s I_a = (2.8\angle 90°)(2092\angle -28.4)$

$= 5858\angle 61.6°$

$E_{g1} = V_A - jX_s I_A = 5181 - j5153$

$|E_{g1}| = 7307$; $|E_{g2}| = 1.1|E_{g1}| = 8038$

$|E_{g1}| \sin\delta_1 = |E_{g2}| \sin\delta_2$

$\sin\delta_2 = \frac{E_{g1}}{E_{g2}} \sin\delta_1 = \frac{-5153}{8038}$

$= -.6411$

$\cos\delta_2 = .7675$

$E_{g2} = |E_{g2}|(\cos\delta_2 + j\sin\delta_2) = 6169 - j5153$

$jX_s I_{a2} = V_A - E_{g2} = 5458\angle 70.8°$

$pf = \cos(70.8 - 90) = .944$

CONCENTRATES

1. $E_g = V_t + (I_t + I_f) R_A$

$\frac{E_g}{E_n} = \frac{V_t + I_t R_A}{E_n} + \frac{R_A I_n}{E_n} \frac{I_f}{I_n}$; $E_N = 260$

$V_t = 240$; $I_t = \frac{10^4}{240}$; $R_A = 1$; $I_n = 2.75$;

$\frac{E_g}{E_n} = 1.0833 + 0.0106 \frac{I_f}{I_n}$

at $\frac{I_f}{I_n} = 1$, $\frac{E_g}{E_n} = 1.094$ (above curve)

at $\frac{I_f}{I_n} = 1.25$, $\frac{E_g}{E_n} = 1.097$ (below curve)

so this line intersects the curve (fig. 7.2) at $\frac{E_g}{E_n} \approx 1.095$. where $\frac{I_f}{I_n} = 1.15$

Then $R_f = \frac{V_t}{I_f} = \frac{240}{1.15 I_n} = 75.8$ ohms

2. Assume that the self-excitation results in a full-load voltage of 240 volts (it doesn't, as the value of R_f is too high, see problem 1).

At no load ($I_L = 0$)

$E_g = I_f (R_A + R_f) = 78 I_f$

normalizing:

$\frac{E_g}{E_n} = 78 \frac{I_f}{I_n} \frac{I_n}{E_n} = 78 \times \frac{2.75}{260} \frac{I_f}{I_n} \times .825 \frac{I_f}{I_n}$

this straight line through the origin intersects the curve of fig 7.2 at apx. $\frac{E_g}{E_n} = 1.2$, $\frac{I_f}{I_n} = 1.45$

or $E_g = 1.2 \times 260 = 312$, $I_f = 3.99$

the $V_L = 312 - 3.99 R_A = 308$ volts

then

$\%\ \text{Regulation} = \frac{308 - 240}{240} \times 100 = 28.3\%$

note: for $R_f = 75.8$, $\frac{E_g}{E_n} = 1.2$, so the regulation is as calculated.

3. $I_f = \frac{200}{100} = 2A \therefore I_A = 23\ A$

$P_{in} = I_f^2 R_f + I_A^2 R_A + P_{rot} + P_{out}$

$P_{rot} = 5000 - 400 - 23^2 \times .5 - 5 \times 746 \frac{W}{HP}$

$= 585$ watts (a)

(b) The total shaft power includes load and rotation losses

$P_{shaft} = 585 + 5 \times 746 = 4315\ W$

$T_{shaft} = K\phi I_A = \frac{4315\ W}{\Omega}$ @ 1000 rpm

$\Omega = \frac{\pi}{30} n = \frac{1000}{30}\pi$ @ 1000 rpm

$K\phi$ = constant

$K\phi I_A = \frac{4315 \times 30}{1000\pi}$ @ $I_A = 23\ A$

$K\phi = 1.7915$

$V_s = .5 I_A + E_g = .5 I_A + K\phi\Omega$

$= .5 I_A + 1.7915 \frac{\pi}{30} n = .5 I_A + .1876 n$

$T = 1.7915\ I_A$

ASSUME rotation losses = 585 W = const at no load

$T\Omega = 585$

$T = \frac{585}{\Omega} = \frac{585}{\frac{\pi n}{30}} = \frac{5586}{n}$

$I_a = \frac{T}{1.7915} = \frac{3118}{n}$

then

$V_s = 200 = .5 \frac{3118}{n} + .1876 n$

or

$.1876 n^2 - 200 n + 1559 = 0$

solving

$n = 533 \pm \sqrt{533^2 - 8310}$

$= 533 \pm 525$

$= 1058 ; 8$

speed increases

$n_{N.L.} = 1058$ RPM

4. $I_o = \frac{500}{.08} = 6250\ A$

at 900 RPM

$T_{900}\Omega = T \times 900 \times \frac{\pi}{30} = 200\ HP \times 746 \frac{W}{HP}$

$T_{900} = 1583$ N-m

$T_{900} = K\phi I_A = K' I_A^2 \quad K' = \frac{T_{900}}{330^2}$

$T_o = \frac{T_{900}}{I_{900}} \times I_o = 1583 \times \left(\frac{6250}{330}\right)^2$

$= 568{,}000$ N-m

5. The parameters given are the per-phase values with the machine in Δ.

$T_{ST}\Omega_s = 3|V_{an}|^2 \frac{a^2 r_r}{(r_s + a^2 r_r)^2 + x_e^2}$ (7.78)

$\Omega_s = 1200 \times \frac{\pi}{30} = 40 \times \pi$

$|V_{an}| = 440/\sqrt{3}$, $R_e = r_s + a^2 r_r$

$.283 = .126 + a^2 r_r$

(a)

$T_{st} = 3 \frac{440^2}{\sqrt{3}^2} \frac{.283 - .126}{.283^2 + .712^2} \frac{1}{40 \times \pi} = 412$ NM

$I_{st} = \frac{440/\sqrt{3}}{(.283^2 + .712^2)^{1/2}} = 332\ A$

(b)

For a Y start, impedances increase by a factor of 3

$T_{st} = 412 \times \frac{3}{3^2} = 137$ NM

$I_{st} = \frac{332}{3} = 111\ A$

6. As in example 7.13

$V_{tap} = 150\left[\left[\frac{.231 - .135}{s} + .135\right]^2 + .934^2\right]^{1/2}$

$s_{switching} = \frac{.096}{-.135 + \left[\left(\frac{V_{tap}}{100}\right)^2 - .934^2\right]^{1/2}}$

STARTING: s = 1

$V_{tap} = 144.3 \qquad s_{sw} = .0995 \rightarrow .1$

6. continued

at $s = .1$

$V_{tap} = 215.88 \rightarrow 216$

$s_{sw} = .053 \rightarrow .06$

at $s = .06$

$V_{tap} = 295 \rightarrow \frac{440}{\sqrt{3}} = 254$

$n_{sw} = 1200(1 - s_{sw})$

n	V_{tap}	n_{sw}
0	144.3	$.9 \times 1200$
1080	216	$.94 \times 1200$
1128	254	

7. NO LOAD, min I_A ∴ $jX_sI_A \perp V_A$

∴ I_A in phase with V_A

$P = V_A I_A = 254\, I_A = 900/3$

$I_A = 1.18$ A

$E_g^2 = (X_s I_A)^2 + V_A^2 = 255^2$

under load $I_A = 12$, $X_s I_A = 240$

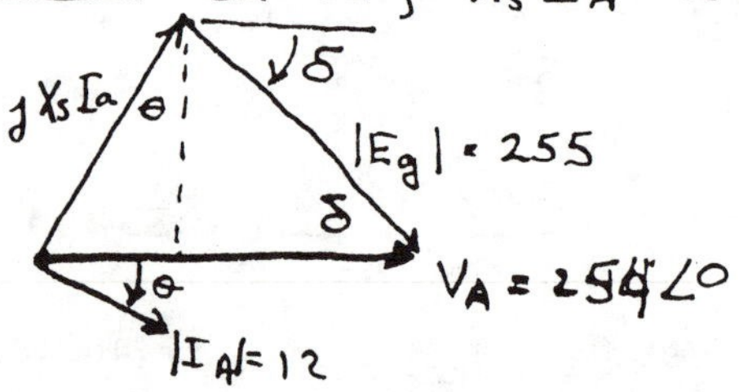

$-X_s I_A \sin\theta + E_g \cos\delta = V_A$ ①

$X_s I_A \cos\theta + E_g \sin\delta = 0$ ②.

SOLVE, using $\sin^2\theta + \cos^2\theta = 1$

$$\cos\delta = \frac{V_A^2 + E_g^2 - X_s^2 I_A^2}{2|V_A||E_g|} = .555$$

(a) $\delta = -56.3°$

(b) solve from ① or ② $\theta = -28°$

$P_{in} = 3 V_A I_A \cos\theta = 8074$ W

$P_{rot} = 900$; $3 I_A^2 R_A = 216$

(c) $P_L = 3(V_A I_A \cos\theta - I_A^2 R_A) - P_{rot}$

$P_L = 6958$ W

(d) $\eta = \frac{P_L}{P_{in}} \times 100 = 86\%$

For 5KVAR P.f. correction

$S = P - jQ = 8074 - j5000 = 3 V_A I_A^*$

$= 3 V_A |I_A| (\cos\theta - j\sin\theta)$

$I_A^* = \frac{8074 - j5000}{3 \times 254} = 10.6 - j6.56$

$jX_s I_A = j20(10.6 + j6.56) = -131.2 + j212$

$E_g = V_A - jX_s I_A = 385.2 - j212$

$|E_g| = 440$, $E_g = k_r I_r$

$k_r = \frac{255}{5} = 51$ $I_r = \frac{440}{51} = 8.63$ A (e)

TIMED

1. Begin at normalization point:

$E_g = 260$, $MMF = 2.75$

Measure slope here (fig. 7.2) as $.71 = OPS$

then $\frac{N_s}{N_f} = \frac{.48}{.71} \times \frac{2.75}{260} = .00725$ (eq. 7.20)

$\left(I_f + \frac{N_s}{N_f} 41.667\right) = 2.75$: $I_f = 2.448$

then

$E_g = 240 + .48(2.448 + 41.667) = 261.2$

$\frac{E_g}{E_n} = 1.0046$

Estimate new MMF_n from slope:

$MMF_n = 1 + OPS(1.0046 - 1) = 1.0033$

slope at this point is slightly lower, say .69. Then $\frac{N_s}{N_f} = \frac{.71}{.69}(.00725) = .0075$

$\left(I_f + \frac{N_s}{N_f} 41.667\right) = (1.0033)2.75$ $I_f = 2.457$

then

$E_g = 240 + .48(2.457 + 41.667) = 261.2$

which is an adequate solution:

$\frac{N_s}{N_f} = .0075$, $R_f = \frac{240}{2.457} = 97.7\,\Omega$

2. $P_{in} = 120V \times 60A = 7200$ W

$P_{out} = 7.5\,HP \times \frac{746\,W}{HP} = 5595$ W

$P_{in} - P_{out} = P_{LOSSES} = 1605$ W

It is necessary to assign these losses to $I_A^2 R_A$, V_A^2/R_f and rotational losses. Here they are divided equally:

$535 = I_A^2 R_A = 60^2 R_A$: $R_A \approx .149\Omega$

$535 = V_A^2/R_f = 120^2/R_f$: $R_f \approx 26.9\Omega$

At this point recalculate and assign remaining losses to rotation

$P_{rot} = 1605 - 60^2 \times .149 - 120^2/26.9$

$= 533$ W

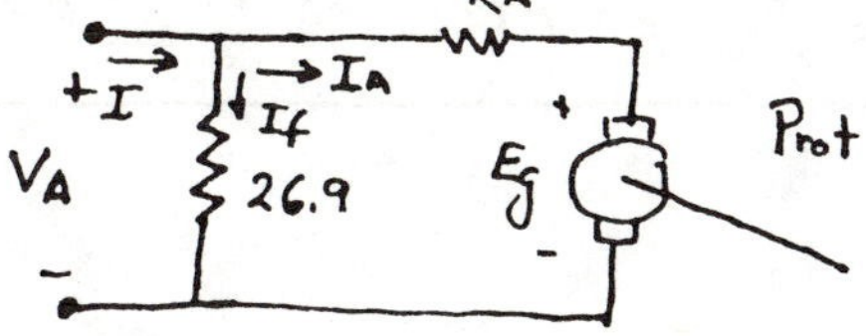

RECALCULATE R_A, with

$I_A = 60 - \frac{120}{26.9} = 55.54$

$R_A = \frac{535}{55.54^2} = .173\Omega$

LOAD TORQUE IS CONSTANT, SO

$P_L = K_T n$: $K_T = \frac{5595}{1000} = 5.595$

Assume Rotation loss is constant

$P_{shaft} = 533 + 5.595\,n$ ①

speed is changed with R_f, V_A is 120.

$E_g = K\phi\Omega$, assume $\Phi \propto I_f = \frac{120}{R_f}$

$E_g = K_V \frac{n}{R_f}$: at 1000 RPM $E_g = 120 - (55.54 \times .173)$ and $R_f = 26.9$

then $K_V = 2.9695$

$P_{shaft} = E_g I_A$

$I_A = \frac{120 - E_g}{.172} = 693.64 - 17.165\frac{n}{R_f}$

$P_{shaft} = 2059.8\left(\frac{n}{R_f}\right) - 50.77\left(\frac{n}{R_f}\right)^2$ ②

EQUATING ① & ② and solving for $\left(\frac{n}{R_f}\right)$ using QUADRADIC EQUATION, then solving for R_f:

$$R_f = \frac{n}{20.205 + \sqrt{20.205^2 - .1098n - 10.5}}$$

at 1150 RPM

$R_f = 31.4\Omega$ (a)

at 750 RPM

$R_f = 19.8\Omega$

3. See fig. 7.7 and example 7.6 on starting $E_g = 0$, $R_1 = \frac{250V}{2.5(125A)}$

$R_1 = 0.8\Omega$.

When I_A drops to 125A

$E_g = 250 - 125(.8) = 150$

and $V_A = E_g + .15 I_A = 168.75$

so CR1 closes at $V_A = 168.75$ V

then $R_2 = \frac{250 - 150}{2.5(125)}$: $R_2 = .32\Omega$

when I_A drops to 125A

$E_g = 250 - 125(.32) = 210$

$V_A = 210 + 125(.15) = 228.75$

so CR2 closes at $V_A = 228.75$

$R_3 = \frac{250 - 210}{2.5(125)} = .128\Omega < R_A$

so no 3rd resistance needed

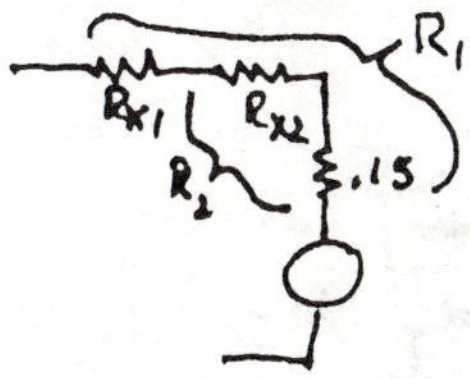

$R_{x2} = .32 - .15 = .17\Omega$

$R_{x1} = .8 - .32 = .48\Omega$

CR1 closes at 169 V

CR2 closes at 229 V

4. NO LOAD: $pf = \frac{P/phase}{V_{l.n.} \times I}$

$pf = \frac{4500/3}{(460/\sqrt{3})\,135} = .0418, \quad \angle I = -87.6°$

at 460 V: $I_{N.L.} = 135\angle -87.6°$

BLOCKED ROTOR:

$|Z_e| = |R_e + jX_e| = \frac{78/\sqrt{3}}{240} = .1876$

eq 7.78

$$a^2 r_r = \frac{(T/3) \times (\pi/30) n_s \, |R_e + jX_e|^2}{V_{an}^2}$$

$T = 17.2 \text{ ft-lbs} \times \frac{746 \text{ N-M}}{550 \text{ ft-lbs}}$

$n_s = 1800$ RPM

$a^2 r_r = .0254$

$\therefore \; R_e = r_s + a^2 r_r \approx 2a^2 r_r \doteq .0508$

then $X_e \cong \sqrt{.1876^2 - .0508^2} = .1806$

then $Z_e \cong .1876\angle 74.2$

At 80% voltage

starting $I = .8\, I_{NL} + \frac{.8 \times 460/\sqrt{3}}{Z_e}$

$I_{start} = 1237\angle -75.4°$

from eq 7.78

$$T_{start}(80\%) = T_{bl.rot} \times \left(\frac{.8\, V_{ln}}{V_{bl.rot}}\right)^2$$

$$= 17.2 \times \left(\frac{.8 \times 460}{78}\right)^2 = 383 \text{ ft-lbs}$$

5.

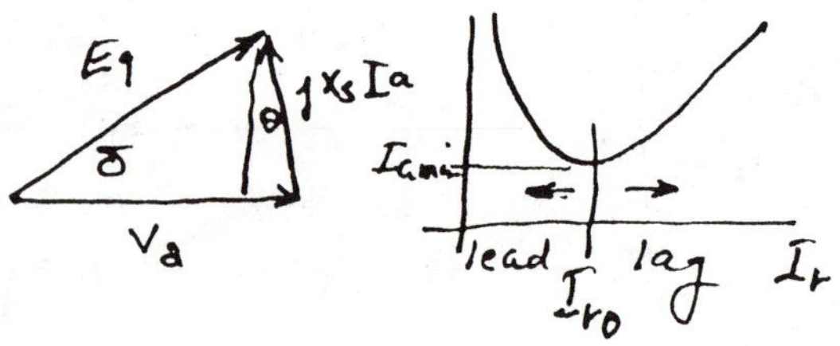

$E_g \sin\delta = X_s I_a \cos\theta = \text{constant}$

$E_g \sin\delta = X_s I_{a\,min}$

$\cos\theta = \frac{I_{a\,min}}{I_a}$ lags for $I_r > I_{ro}$

$I_{a\,min} = 15$

$I_a = \frac{15}{\cos\theta}$

(a) 80% lag: $I_a = \frac{15}{.8} = 18.75$ A

interpolate table

$\frac{I_r - 15}{18.75 - 18} = \frac{20 - 15}{32 - 18}; \quad I_r = 15.27$ A

(b) Table (left side)

$\frac{I_r - 3}{40 - 45} = \frac{5.5 - 3}{30 - 45} \quad I_r = 3.46$ A

(c) $\cos\theta = \frac{15}{40} = $ p.f. $= .375$ leading

WARM UPS

1. from eq 8.8

$$\frac{V_o}{V_{in}} = \frac{\frac{-25\times10^3}{5\times10^3}}{S/\omega_{cf} + 1}$$

from eq 8.9 $\omega_{cf} = \dfrac{\omega_o \times 5K}{5K + 25K} = \dfrac{\omega_o}{6}$

so

$$\frac{V_o}{V_{in}} = \frac{-5}{\frac{S}{2\pi/6}\times10^{-7} + 1}$$

the gain is -5, $BW = 1.67\ MHz$

2. the voltage at the N terminal is:

$$V_n = \frac{R_2}{10K + R_2} V_o + \frac{S}{\omega_o} V_o$$

at low frequency

$$V_n = \frac{R_2}{10K + R_2} V_o = \frac{R_2}{10K + R_2}(V_1 + V_2)$$

as specified.

Using voltage division at the inputs:

$$\frac{V_1 R_1}{R_1 + 10K} + \frac{V_2\, 10K}{R_1 + 10K} = V_n$$

matching coefficients of V_1.

$$\frac{R_1}{R_1 + 10K} = \frac{R_2}{R_2 + 10K} \quad (1)$$

matching coefficients of V_2

$$\frac{10K}{R_1 + 10K} = \frac{R_2}{R_2 + 10K} \quad (2)$$

as (1) and (2) are equal $R_1 = 10K$

then $\dfrac{R_2}{R_2 + 10K} = \dfrac{1}{2}$, $R_2 = 10\ K$

3. from eq. 8.25

$$\frac{V_o}{V_{in}} = \frac{-R_{fb}/R_{in}}{S\, C_{fb} R_{fb} + 1} = \frac{-10}{\frac{S}{60\pi} + 1}$$

from specifications:

$$C_{fb} R_{fb} = \frac{1}{2\pi \times 30} \quad \text{and } C_{fb} = 10^{-7}$$

then $R_{fb} = 53.05\ K\Omega$

as $\dfrac{R_{fb}}{R_{in}} = 10$ $\therefore$ $R_{in} = 5.305\ K\Omega$

4. see eq. 4.85

$$\left(\frac{S}{\omega_o}\right)^2 + \frac{1}{Q}\left(\frac{S}{\omega_o}\right) + 1 \quad ; Q = .5$$

match with eq. 8.32 denominator with $\omega_o = 100$

$$R_2 R_5 C_3 C_4 = \frac{1}{100^2} \quad ; R_2(C_1 + C_3 + C_4) = \frac{1}{50}$$

with $R_2 = 10K\Omega$

$$C_1 + C_3 + C_4 = 2\times10^{-6} \quad (1)$$

$$R_5 C_3 C_4 = 10^{-8} \quad (2)$$

The specification on capacitances cannot be met, so that is relaxed to allow $1\mu F$ capacitance then make up two of the capacitances with two $1\mu F$ capacitors in series, say C_3 & C_4 and use a single capacitor for $C_1 = 1\mu F$.

Then from (2) $R_5 = 40\ K\Omega$

5. Equation 8.48

$$I_D(mA) = 5\left[1 - \frac{-3}{-5}\right]^2 = .8\ mA$$

note that $V_{gs\,(off)} = -V_{po} = -5$ volts

6. Kirchhoff's Voltage law

$$V_{DD} = I_D R_D + V_{DS} + I_D R_S$$

$$30 = .005 R_D + 15 + .005 R_S$$

then $R_D + R_S = 3000\ \Omega$

From characteristic given:

$$V_{gs} \approx -1.8 \text{ volts}$$

then $I_D R_S = 1.8$ volts

$$R_S = \frac{1.8}{.005} = 360\ \Omega$$

then $R_D = 3000 - 360 = 2{,}640\ \Omega$

7. $V_{GS} = 3$ will make the operating point near the middle of the linear region. This causes $I_d = 2.5 mA$, so $V_s = 2.5$ volts. The bias circuit is as figure 8.37(a) with $K = \frac{3}{16}$

8. INPUT CIRCUIT

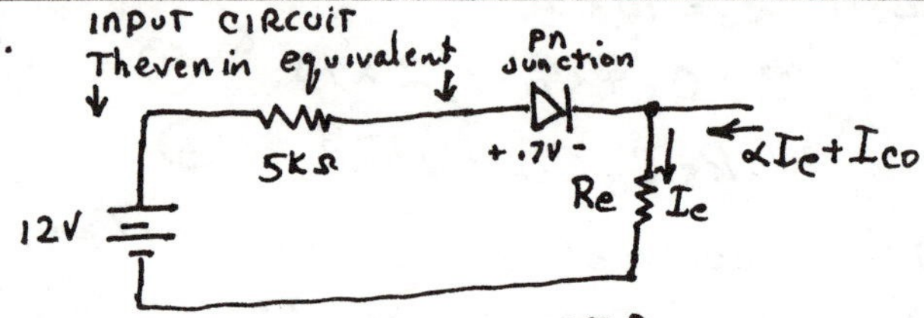

$$I_C = \alpha I_e + I_{co} = .015 A$$

then $.95 I_e + 10^{-6} = .015$; $I_e = 15.8 mA$

Kirchhoff's Voltage Law

$$12 = 5000 I_b + .7 + .0158 R_E$$

$$I_b = I_e - I_c = .0008$$

$$R_e = 462\ \Omega \quad (a)$$

$$S_v = \frac{2.5\alpha_o}{R_b(1-\alpha_o) + R_e} = \frac{2.5 \times .95}{5K(.05) + 462} = .00334$$

(b)

$$\Delta I_c = S_v \Delta T = .00334 \times 125 = .418 \text{ mA}$$

9. $$\left.\frac{\Delta V_{CE}}{\Delta I_c}\right|_{I_b = 1mA} = \frac{40-0}{.1-.08} = 2000\ \Omega = r_c$$

$$\left.\frac{\Delta I_c}{\Delta I_b}\right|_{V_{CE} = 10} = \frac{.145 - .015}{.002 - 0} = 75 = \beta$$

10. $$i_b = \frac{V_{in}}{h_{ie}}$$

$$V_{out} = -\beta i_b (r_c \| R_c \| R_L)$$

$$\frac{V_{out}}{V_{in}} = \frac{-\beta(r_c \| R_c \| R_L)}{h_{ie}} = -41.7$$

11. $P_{out\,(noise)} = N_{out} = 4kTR_g B \times FG = 10^{-12}$

Assume $R_S = R_{in} = 50\ \Omega$

$$F_{db} = 10 \log F = 6 \therefore F = 3.98$$

$$G = \frac{10^{-12}}{4 \times 1.38 \times 10^{-23} \times 293 \times 50 \times 10^5 \times 3.98}$$

(a) $G = 3.11$

(b) $T_e = 290(F-1) = 864^\circ$ (eq. 8.193)

(c) $R_e = 50(F-1) = 149\ \Omega$ (eq. 8.195)

CONCENTRATES

1. Begin with the left most source and convert to thevenin equivalent

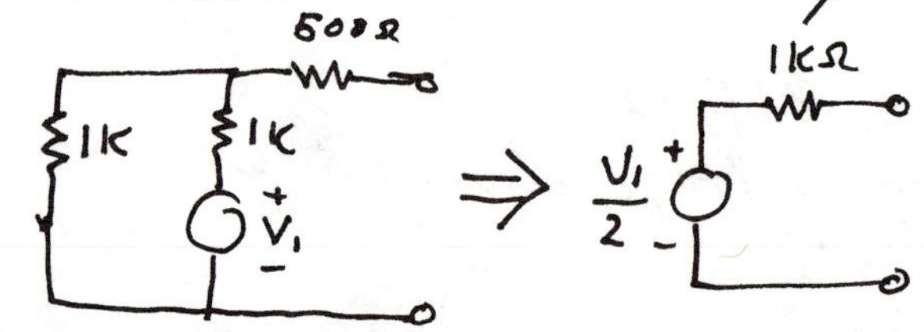

Combine in parallel with next source

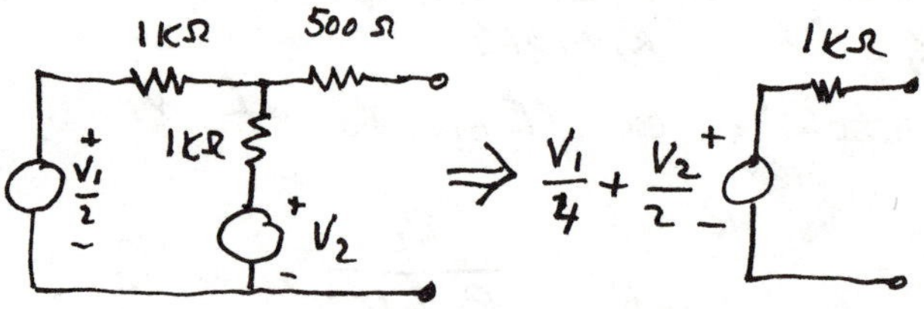

Continue two more reductions:

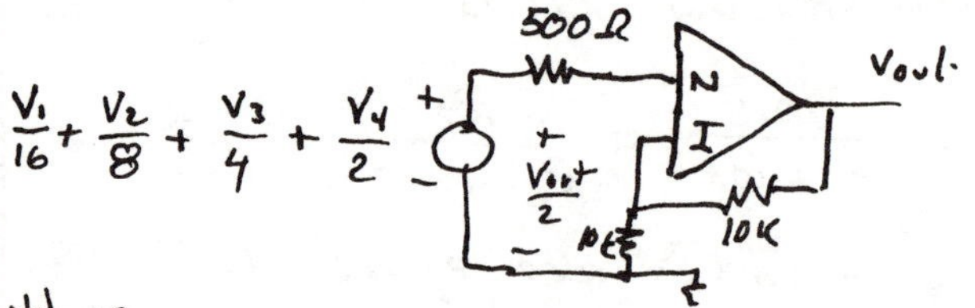

thus

$$V_{out} = \frac{V_1}{8} + \frac{V_2}{4} + \frac{V_3}{2} + V_4$$

2. $e_{out} \approx -\dfrac{Z_{fb}}{Z_{in}} v_s$

$$Z_{fb} = \frac{1}{j\omega C_x} = \frac{.225d}{j\omega A}$$

$$Z_{in} = \frac{1}{j\omega C}$$

$$e_{out} = -\frac{.225 C d}{A} v_s$$

3. Refer to fig. 8.39 : r_{ds} and C_{ds} are negligible as they have not been given.

Find the Miller effective capacitance

$$C_m = C_{oss}(1 + g_m R_o)$$

$$= 2pF(1 + 2\times10^{-3}\times10^4) = 42\,pF$$

Then $C_{gs} = C_m + (C_{iss} - C_{oss}) = 46\,pF$

All capacitances shown in the problem circuit are shorted at high frequency. The frequency response then follow v_{gs}:

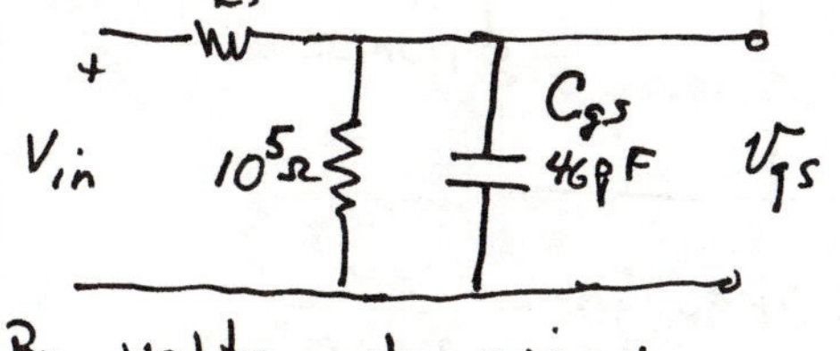

By voltage division:

$$\frac{V_{gs}}{V_{in}} = \frac{10^5}{R_s + 10^5} \times \frac{1}{\frac{10^5 R_s}{10^5 + R_s} C_{gs} s + 1}$$

as $R_s << 10^5$

$$\omega_{c.o.} \cong \frac{1}{R_s C_{gs}} \qquad f_{c.o.} = \frac{1}{2\pi R_s C_{gs}}$$

a.) $R_s = 5K$ $\quad f_{c.o.} = 692$ kHz

b) $R_s = 50$ $\quad f_{c.o.} = 69.2$ MHz

4. $G_1 = 12db \rightarrow 15.9 \quad F_1 = 6db \rightarrow 4.0$

$G_2 = -6db \rightarrow .25 \quad F_2 = 12db \rightarrow 15.9$

$G_3 = 40db \rightarrow 10^4 \quad F_3 = 6db \rightarrow 4.0$

(a) $F = F_1 + \dfrac{F_2 - 1}{G_1} + \dfrac{F_3 - 1}{G_1 G_2} = 5.69$ or 7.6 db

(b) $\dfrac{S_{out}}{N_{out}} = 1 = \dfrac{G_1 G_2 G_3 S_{in}}{F G_1 G_2 G_3 N_{in}} = \dfrac{S_{in}}{F N_{in}}$

note: Bandwidth for noise set by I.F. stage

$N_{in} = 4kTR_{in}B$; assume $R_{in} = 72\,\Omega$

$s_{in} = F N_{in} = 5.69\,(4\times1.38\times10^{-23}\times293\times72\times10^{7})$

$= 6.6\times10^{-11}$ W

5. Small-signal equivalent circuit

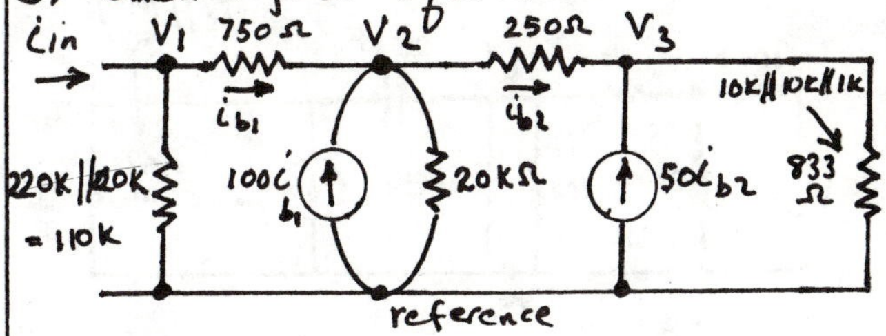

$$i_{b1} = \frac{V_1 - V_2}{750}, \quad i_{b2} = \frac{V_2 - V_3}{250}$$

NODE EQUATIONS (Kirchhoff Current Law)

$$i_{in} = V_1\left(\frac{1}{750} + \frac{1}{1.1\times10^4}\right) - V_2\left(\frac{1}{750}\right)$$

$$100\frac{V_1 - V_2}{750} + \frac{V_1 - V_2}{750} = \frac{V_2}{20K} + \frac{V_2 - V_3}{250}$$

$$\frac{V_2 - V_3}{250} + 50\frac{V_2 - V_3}{250} = \frac{V_3}{833}$$

solve for V_3

$$V_3 = 10^5 I_{in}$$

$$I_L = \frac{V_3}{1000} = 100 I_{in}, \quad A_I = 100$$

6. With C_1 shorted, the impedance seen at the base is $6K||3K||1K$ or $667\,\Omega$. The emitter output impedance is then

$$\frac{667 + h_{ie}}{1 + \beta} = 29.2\,\Omega$$

(continues)

6. continued

C_E then "sees" $29.9 \| 200 = 25.5\,\Omega$ so the low-frequency "corner" due to C_E is

(a) $f_1 = \dfrac{1}{2\pi C_E \times 25.5} = 62.4$ Hz

The worst case effect on C_1 is with C_E shorted, in which case C_1 "sees" $1\,K\Omega$ in series with the parallel combination of $6K \| 3K \| h_{ie} = 400\,\Omega$. the

(b) $C_1 \geq \dfrac{1}{2\pi \times 10 \times 1400} = 11.4\,\mu F$

HIGH FREQUENCY CIRCUIT:

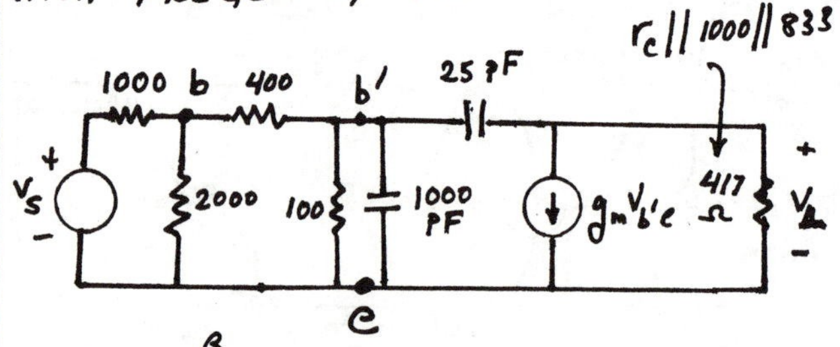

$g_m = \dfrac{\beta}{r_{b'e}} = .39$

Thevenin equivalent left of b'e:

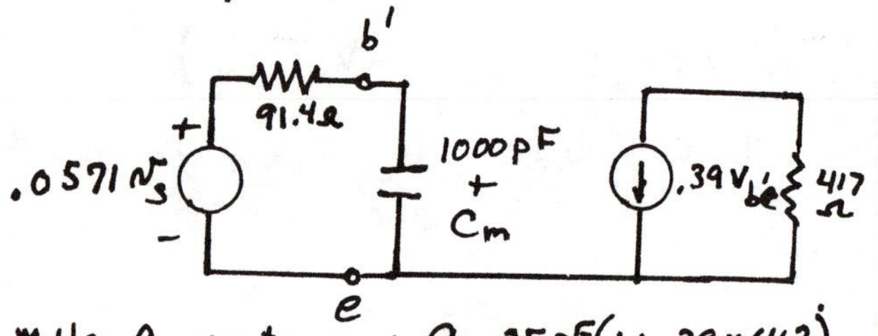

Miller Capacitance : $C_m = 25\,pF(1 + .39 \times 417)$
$= 4091\,pF$

$V_L = -.39 \times 417\, V_{b'e}$

$V_{b'e} = \dfrac{.0571\, V_s}{j\omega\, 91.4\, C + 1}$ $C = 5091 \times 10^{-12} F$

$\dfrac{V_L}{V_s} = \dfrac{9.29}{j\dfrac{\omega}{2.15 \times 10^6} + 1}$ $\therefore f_2 = 342\,KHz$

$A_{db} = 20 \log 9.29 = 19.4$ db

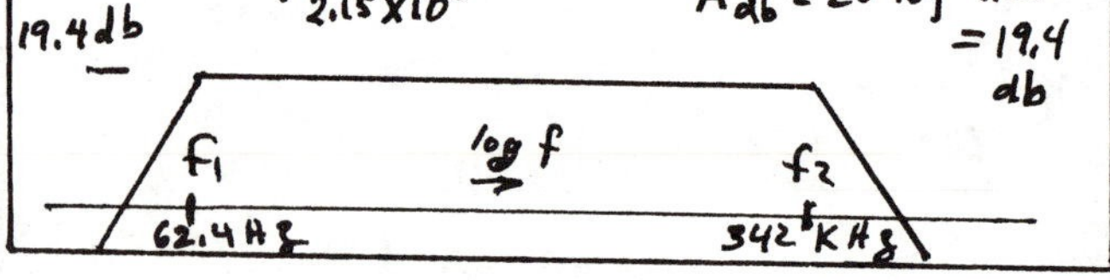

7. As fig. 8.36 for voltage feedback biasing, plot $V_{gs} = V_{ds}$ curve, find Q point at intersection with load line at $V_{dsQ} = -13$ volts, $I_{dQ} = -14.5\,mA$

Obtain r_{ds} from the average of the characteristic curves slopes for $V_{gs} = -12$ and $V_{gs} = -14$:

$r_{ds} = \dfrac{4.2K\Omega + 3.1K\Omega}{2} = 3.65\,K\Omega$

At $V_{ds} = -13$: $I_d\big|_{V_{gs}=-12} - I_d\big|_{V_{gs}=-14} = 4.5\,mA$

$g_m = \dfrac{4.5\,mA}{2\,V} = 2.25 \times 10^{-3}$ mhos

At low frequencies, the time constant is set by the 100K resistances and the two C's, with some feedback effect. Ignoring the feedback, using voltage division:

$V_{gs} \approx V_{in} \dfrac{R + 1/sC}{R + 1/sC + 1/sC} = V_{in} \dfrac{sCR+1}{sCR+2}$

so $f_1 \approx \dfrac{1}{2\pi} \dfrac{2}{RC} = 31.8$ Hz

However, the output capacitance see's apx 1600 ohms, so the corner frequency associated with the output circuit is

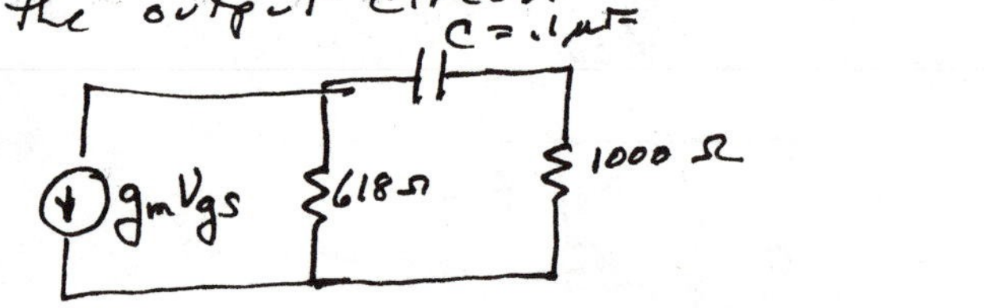

$V_{out} \cong -618\, g_m V_{gs} \dfrac{1000}{1618 + \dfrac{10^7}{s}}$

$= \dfrac{-1.39 \times 10^{-4} s}{\dfrac{s}{6180} + 1}$

from the output capacitance

$f_1 = \dfrac{6180}{2\pi} = 984$ Hz

So the output capacitance dominates the lower cuttoff frequency

(CONTINUES)

7. continued

The high frequency circuit includes the capacitances as 8.39, with all other C's shorted. The Miller effective capacitance is found:

$$C_m = C_{oss}(1+g_m R_o) = 2.5(1+.00225 \times 382)\,pF$$
$$= 4.6\ pF$$

the equivalent circuit is then

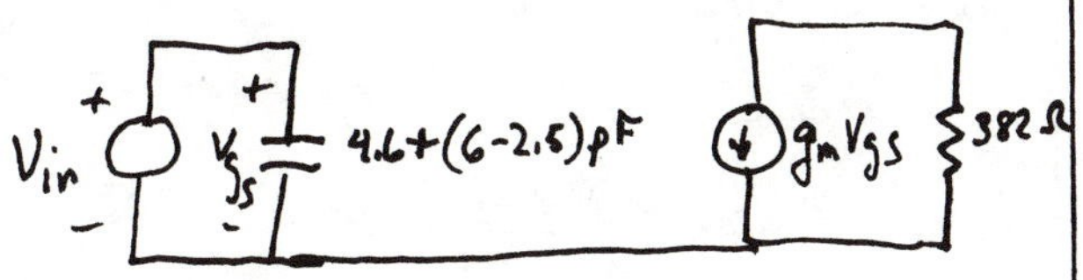

$$V_{out} = -382\, g_m V_{gs} = -.859\ V_{in}$$

because there is no source resistence there is no upper frequency limit.

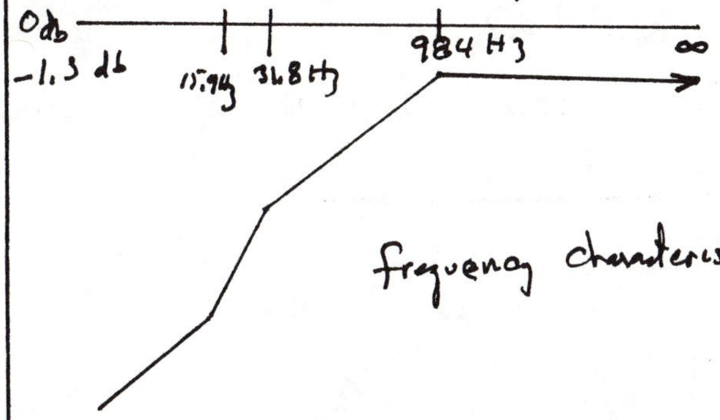

frequency characteristic

8. from specifications and using Kirchhoff's voltage law:

$$V_S = 15 - 600 I_{DQ} - 5 \doteq 4 \text{ volts}$$

$$V_S = I_{dQ} R_s \quad \therefore R_s = 400\,\Omega$$

from Q point: $V_{gs} = -3.4 \therefore V_g = .6\,V$

$$V_g = \frac{R_2}{R_1+R_2} \times 15 \ ; \ \frac{R_1 R_2}{R_1+R_2} = 10^5 \therefore V_g = \frac{15 \times 10^5}{R_1}$$

then $R_1 = \frac{15 \times 10^5}{.6} = 2.5 \times 10^6$

$R_2 = 1.04 \times 10^5$

$$V_o = -600\, g_m V_i \therefore A_V = \frac{V_o}{V_i} = -1.32$$

TIMED

1. Using Kirchhoff's current law and defining $k = \frac{R_3}{R_3+1000}$

$$\frac{V_{out}}{V_{in}} = k\,\frac{s^2 C_1 R_1 C_2 R_2 + s\left[(C_1+C_2)R_1 + C_2 R_2 \left(1-\frac{1}{k}\right)\right] + 1}{s^2 C_1 R_1 C_2 R_2 + s(C_1+C_2)R_1 + 1}$$

from specifications:

$$C_1 R_1 C_2 R_2 = \frac{1}{(2\pi \times 400)^2} \quad (1)$$

At 400 Hz,

$$\left.\frac{V_{out}}{V_{in}}\right|_{400 Hz} = k\,\frac{(C_1+C_2)R_1 + C_2 R_2 \left(1-\frac{1}{k}\right)}{(C_1+C_2)R_1}$$

from specifications

$$\left.\frac{V_{out}}{V_{in}}\right|_{400 Hz} \le -30\ db$$

Set $\left.\frac{V_{out}}{V_{in}}\right|_{400 Hz} = 0$

then $\frac{1}{k} = 1 + \frac{(C_1+C_2)R_1}{C_2 R_2}$

for $C_1 = C_2 = 10^{-8}\,F$

$$R_3 = 500\,\frac{R_2}{R_1} \quad (2)$$

from specifications of 3 db point (i.e. -27db) at 400 ± 40 Hz

$$\left.\frac{V_{out}}{V_{in}}\right|_{360 Hz} = \frac{\left(-\left(\frac{360}{400}\right)^2 + 1\right)k}{-\left(\frac{360}{400}\right)^2 + j\,720(C_1+C_2)R_1 + 1} = .0477k$$

from which $(C_1+C_2)R_1 = .0059$

with $C_1 = C_2 = 10^{-8}F$ $\quad R_1 = 295\ K\Omega$

from (1) $\quad R_2 = 5.37\ K\Omega$

from (2) $\quad R_3 = 9.096\ \Omega$

2.

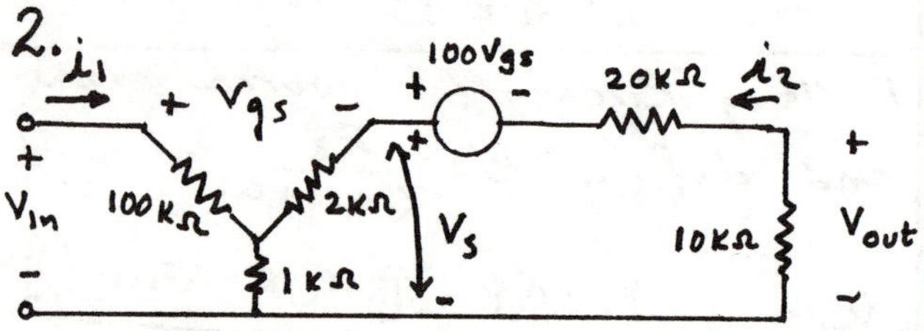

INPUT LOOP: $V_{IN} = 101K i_1 + 1K i_2$ ①

OUTPUT LOOP: $100 V_{gs} = 1K i_1 + 33K i_2$ ②

Also: $V_s = 1K i_1 + 3K i_2$

$V_{gs} = V_{IN} - V_s = 100K i_1 - 2K i_2$ ③

substitute ③ into ②, obtain

$$10,000 i_1 = 233 i_2 \quad ④$$

note that $V_{out} = -10K i_2$ ⑤

from ① and ④ and ⑤

(a) $A_v = \dfrac{V_{out}}{V_{in}} = -2.98$

from ① and ④

(b) $Z_{in} = \dfrac{V_{in}}{i_1} = 144 K\Omega$

3.

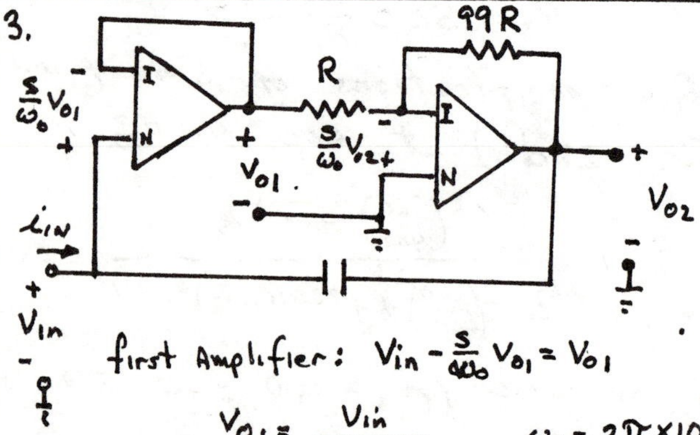

first Amplifier: $V_{in} - \dfrac{S}{\omega_0} V_{01} = V_{01}$

$$V_{01} = \dfrac{V_{in}}{S/\omega_0 + 1} \qquad \omega_0 = 2\pi \times 10^7$$

Second Amplifier: KCL at I terminal

$$\dfrac{V_{01} + \dfrac{S}{\omega_0} V_{02}}{R} + \dfrac{V_{02} + \dfrac{S}{\omega_0} V_{02}}{99R} = 0$$

$$V_{02} = \dfrac{-99 V_{01}}{\dfrac{100 S}{\omega_0} + 1} = \dfrac{-99 V_{in}}{\left(\dfrac{100 S}{\omega_0} + 1\right)\left(\dfrac{S}{\omega_0} + 1\right)}$$

$$i_{in} = SC(V_{in} - V_{02})$$

so

$$i_{IN} = SC V_{in}\left[1 + \dfrac{99}{\left(\dfrac{100S}{\omega_0}+1\right)\left(\dfrac{S}{\omega_0}+1\right)}\right]$$

$$= 100 SC V_{in}\left[\dfrac{\left(\dfrac{S}{\omega_0}\right)^2 + 1.01\dfrac{S}{\omega_0} + 1}{\left(\dfrac{100S}{\omega_0}+1\right)\left(\dfrac{S}{\omega_0}+1\right)}\right]$$

$$Z_{in} = \dfrac{1}{100 SC} \dfrac{\left(\dfrac{100S}{\omega_0}+1\right)\left(\dfrac{S}{\omega_0}+1\right)}{\left(\dfrac{S}{\omega_0}\right)^2 + 1.01\dfrac{S}{\omega_0} + 1}$$

this appears to be a capacitance of $100C$ for frequencies below where $\dfrac{100\omega}{\omega_0} \ll 1$ or

where $\dfrac{100f}{f_0} \ll 1$: $f \ll \dfrac{f_0}{100} = 10^5 Hz$

4.

$= 870 \Omega$

└ bias resistance is redundant for voltage gain calculation

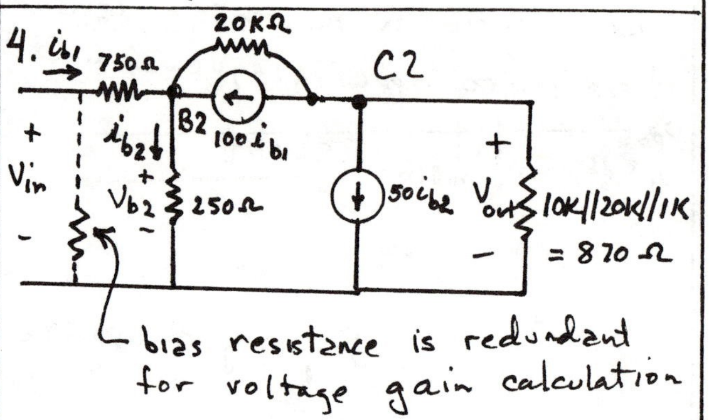

$i_{b_1} = \dfrac{V_{in} - V_{b2}}{750} \qquad i_{b2} = \dfrac{V_{b2}}{250}$

Kirchhoff's current law at B2

$$\dfrac{V_{in} - V_{b2}}{750}(1+100) = \dfrac{V_{b2}}{250} + \dfrac{V_{b2} - V_{out}}{20K}$$

at C2 node

$$100\dfrac{V_{in} - V_{b2}}{750} + 50\dfrac{V_{b2}}{250} + \dfrac{V_{out} - V_{b2}}{20K} + \dfrac{V_{out}}{870} = 0$$

Solve for V_{out}

$$\dfrac{V_{out}}{V_{in}} = A_v = -162$$

WARM UPS

1.

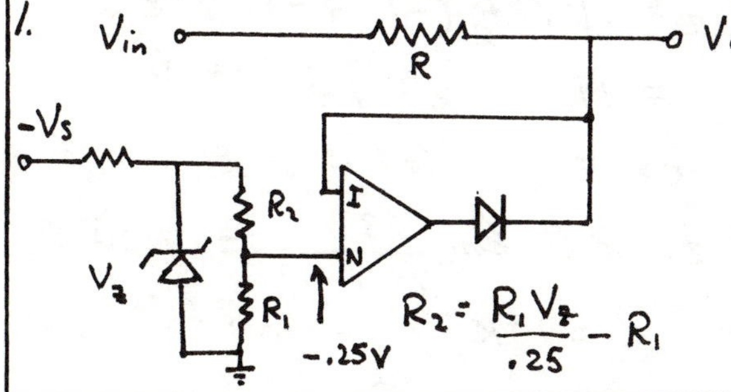

$$R_2 = \frac{R_1 V_z}{.25} - R_1$$

2. reference circuit

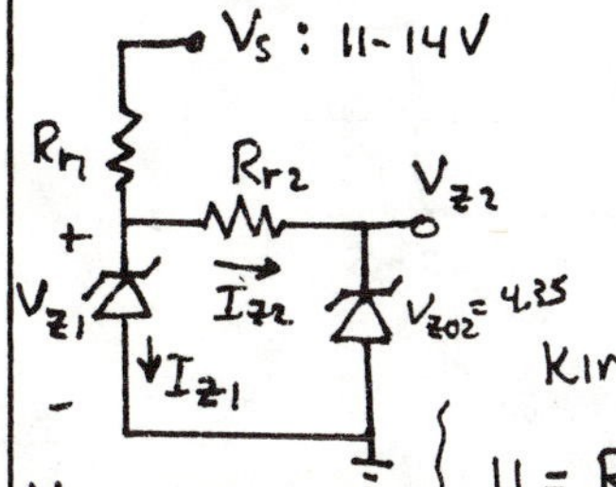

At $V_s = 11$ volts set

$I_{z1} = 8$ mA, $V_{z1} = 8.1$ V

$I_{z2} = 10$ mA, $V_{z2} = 4.7$ V

Kirchhoff's voltage law

$11 = R_{r1}(I_{z1}+I_{z2}) + 8.1$

$8.1 = R_{r2} I_{z2} + 4.7$

$V_{z01} = 8.2 + .008 \times 8$

$V_{z01} = 8.036$

then $R_{r2} = 340\,\Omega$, $R_{r1} = 161\,\Omega$

with $V_s = 14$ Volts

$$14 = (I_{z1}+I_{z2})161 + 8I_{z1} + 8.036$$

$$14 = (I_{z1}+I_{z2})161 + 340 I_{z2} + 35 I_{z2} + 4.35$$

solve for $I_{z2} = .0104$ A

$$V_{z2} = 35 \times .0104 + 4.35 = 4.712$$

$$\text{Regulation} = \frac{4.712 - 4.7}{4.7} \times 100 = 0.25\%$$

3. from the load line with no load, $I_C \approx 4.8$ mA, and $I_b = 4$ mA is adequate to maintain saturation in all cases. As $I_{cbo} = .4$ ma, for 10 loads, the additional I_C is $10 \times .4 = 4$ mA, for a total of 8.8 mA

(a) this requires between .4 and .5 mA of base current: $I_b \approx .45$ mA

note: the scales for I_b are .1, .2, .3 and .4 mA

(b) The maximum base current is .45 ma, so $.45 \times 10 = 4.5$ mA of drive current with the transistor off. This leaves 0.5 volts to drive 4.5 mA, for a parallel load of $\frac{.5\,V}{.0045} = 1111\,\Omega$

each R_b is then $\leq 11.11\,K\Omega$ (b)

4. $T_{min} = \tau_d + \tau_r + \tau_s + \tau_f = 5 + 50 + 30 + 30$ nS

$$f_{max} = \frac{1}{T_{min}} = 8.7\text{ MHz}.$$

5. Voltage Table

V_A	V_B	V_C
H	H	L
H	L	H
L	H	H
L	L	H

TRUTH TABLE

H = 0, L = 1

A	B	C
0	0	1
0	1	0
1	0	0
1	1	0

$C = \overline{A+B}$

6. for a pulse duration of 10μS, The pulse area must excede $50\mu s \times I_{gT}$

$$I_g \times 10\mu S = 50\text{mA} \times 50\mu S$$

$$I_g = 250\text{ mA}$$

7. SWITCHING TAKES place when

$$V_E = \frac{V_{bb} \times 10}{24} = \frac{V_{bb}}{2.4}$$

the time function is

$$V_E = V_{bb}(1 - e^{-t/\tau})$$

$$V_E(\text{switch}) = V_{bb}(1 - e^{-T/\tau}) = \frac{V_{bb}}{2.4}$$

Thus T does not change for changes of V_{bb}.

$$V_P = \frac{V_{bb}}{2.4}$$ so at $V_{bb} = 24(1.25)$

$V_P = 12.5$ V

at $V_{bb} = 24(.75)$

$V_P = 7.5$ V

8. SWITCH OPENS at $t = 0$

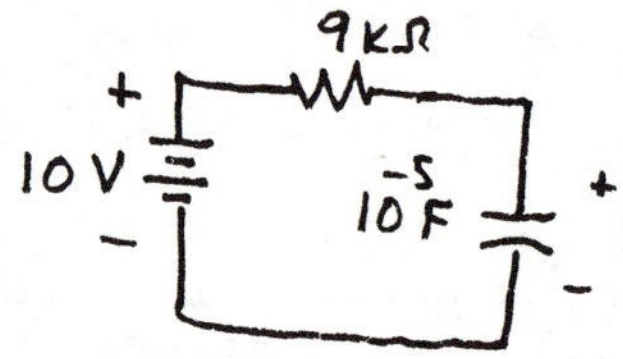

① $V_c(t) = 10 + (V_c(0) - 10)e^{-\frac{t}{.09}}$ $\quad 0 \le t \le .5$

SWITCH CLOSES at $t = .5$

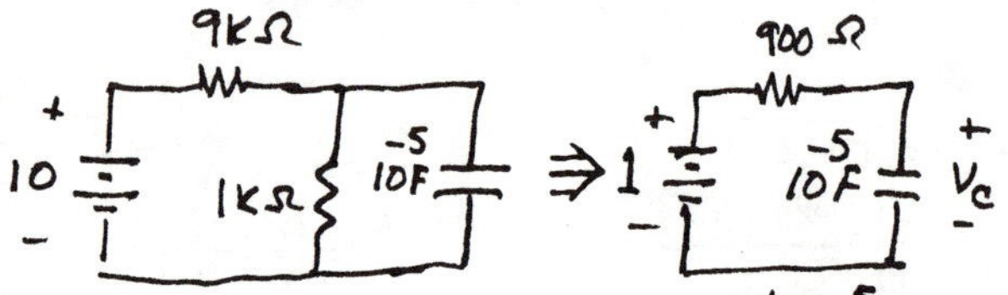

$V_c(t) = 1 + [V_c(.5) - 1]e^{-\frac{t-.5}{.009}}$ ②

$.5 \le t \le 1$

at $t = 1$, for steady state, $V_c(1) = V_c(0)$

$V_c(0) = V_c(1) = 1 + [V_c(.5) - 1]e^{-\frac{.5}{.009}}$

substitute for $V_c(0)$ into ① and obtain: $V_c(.5) = 9.97$ V

substitute this into ② to obtain

$V_c(T) = V_c(0) = 1.00$ V

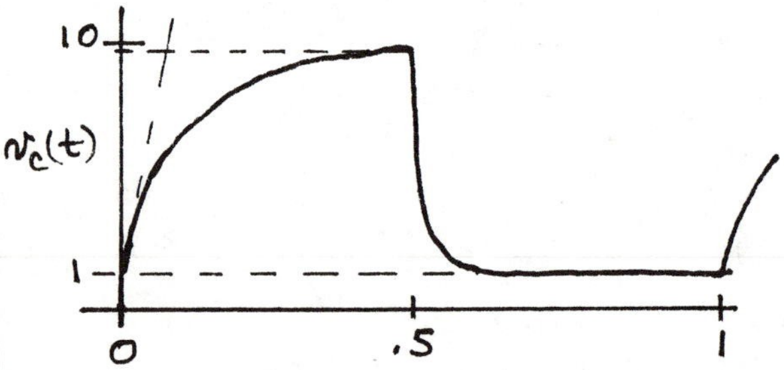

9. In the quiescent state B2 is at .6 V, and C_1 is charged to $-(12 - .6) = -11.4$ V. When T1 is turned on its collector falls to about .3 volts, putting B2 at -11.1 volts. V_{C1} then charges through R_1C_1 toward $(12 - .3)$ V. when it reaches $+.3$ V, $V_{B2} = .6$ and T2 turns on to end the pulse.

$$V_{C2} = 11.7 + (-11.4 - 11.7)e^{-t/R_1C_1}$$

$$= 11.7 - 23.1e^{-t/R_1C_1}$$

at $V_{C2} = .3$ Volts

$$t = R_1C_1 \ln \frac{23.1}{11.7 - .3} = .706 R_1C_1$$

with $R_1 = 10K$, $C_1 = 10^{-7}$ F

$t = .706$ mS

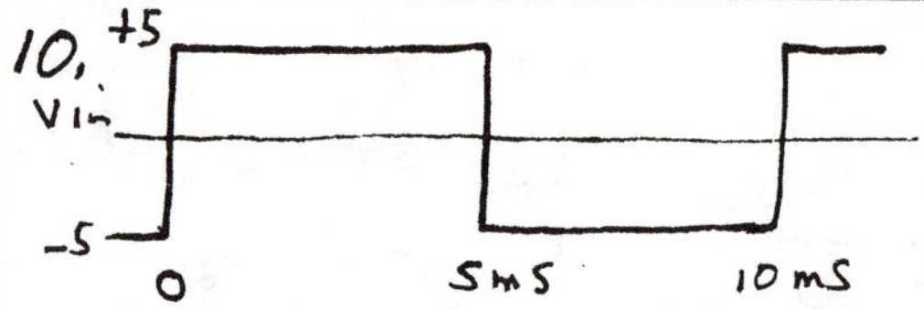

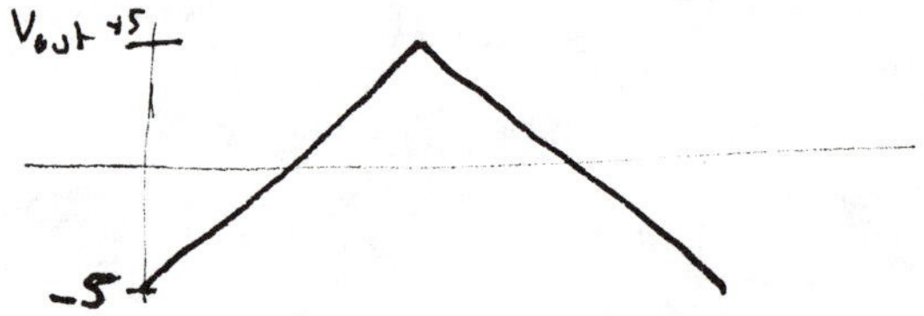

$V_{out} = k\int V_{in}\,dt = k5t + \text{const}$

at $t = 0$ $\quad V_{out} = -5 = \text{const.}$

$V_{out} = 5kt - 5$

at $t = .005$ $\quad V_{out} = 5$

$5k(.005) = 10$; $k = 400$

then $\dfrac{V_{out}}{V_{in}} = \dfrac{400}{s}$

for inverting op-amp:

$$\frac{V_{out}}{V_{in}} = -\frac{Z_{fb}}{Z_{in}}$$

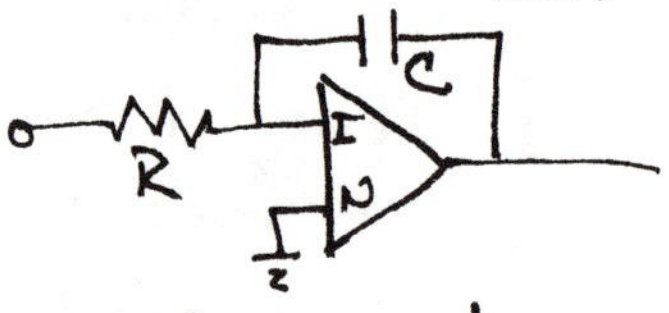

$$\frac{V_{out}}{V_{in}} = -\frac{1}{sCR} \quad \therefore CR = \frac{1}{400}$$

CONCENTRATES

1. With both diodes "open"

$$\frac{V_o}{V_{in}} = -\frac{10K}{10K} = -1 \quad (1)$$

with upper diode conducting and lower diode "open, redraw circuit:

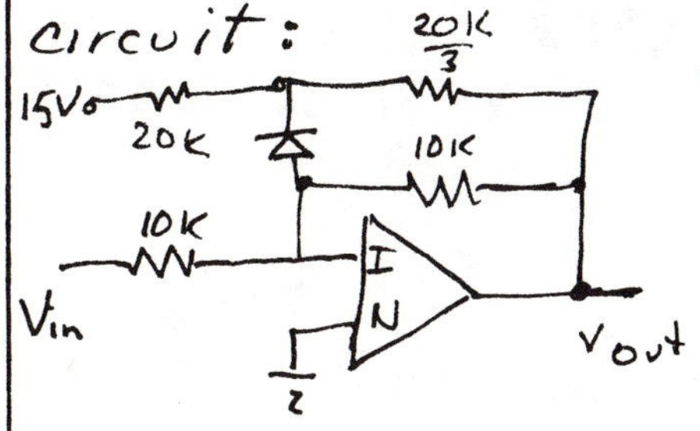

$\frac{20K}{3} \| 10K = 4K$

$$V_{out} = -\frac{4K}{10K}V_{in} - \frac{4K}{20K}15$$

$$= -.4\,V_{in} - 3 \quad (2)$$

with lower diode conducting and upper diode "open":

$$V_{out} = -.4\,V_{in} + 3 \quad (3)$$

Check intersection

(1) & (2) $-V_{in} = -.4\,V_{in} - 3$ $\quad V_{in} = 5$

(1) & (3) $-V_{in} = -.4\,V_{in} + 3$ $\quad V_{in} = -5$

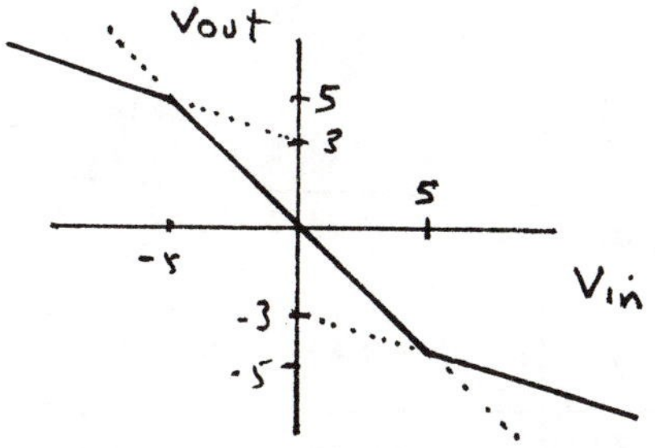

2. Refer to figure 5.9

Assume: $I_{ZK} = 5$ mA

Determine V_{ZO}

$$16 = .155 \times 4 + V_{ZO} : \; V_{ZO} = 15.38$$

Determine R_S from minimum V_S and maximum load current:

$$24 = (.5 + .005)R_S + .005\,R_Z + 15.38$$

then $R_S = 17.03$ ohms

At $V_{S\,MAX} = 30$ V and $I_L = 0$

$$30 = I_Z(17.03 + 4) + 15.38$$

$$I_Z = 695 \text{ mA}$$

The maximum I_Z is given as 530 mA, so this diode is inadequate for the design.

The best design possible is determined by setting the maximum zener current at maximum V_S with $I_L = 0$:

$$30 = .53\,(R_S + 4) + V_{ZO}$$

$$R_S = 23.6 \text{ ohms}$$

This design will fail to regulate when $I_Z < I_{ZK}$. the maximum regulated load current is when $I_Z = I_{ZK}$:

$$V_S = R_S(I_{L\,max} + I_{ZK}) + V_{ZO} + R_Z I_{ZK}$$

$$I_{L\,max} = \frac{V_S - 15.52}{23.6}$$

the worst case is with $V_{S\,MIN} = 24$

$$I_{L\,MAX} = 359 \text{ mA}$$

with $V_{S\,MAX} = 30$ V: $I_{L\,MAX} = 614$ mA

With $V_S = 24$ V and $I_L = 500$ mA the zener is not functioning, so $V_L = 24 - .500\,(23.6) = 12.2$ Volts

With $V_S = 30$ V and $I_L = 0$

$$V_L = V_{ZO} + .005\,R_Z = 15.52 \text{ volts}$$

then

$$\text{Regulation} = \frac{15.52 - 12.2}{12.2} \times 100 = 27.2\%$$

The regulation can be improved by
(1) Using a higher power zener.
(2) Using a pass transistor.
(3) Using an op-Amp regulator.
(2) is cheapest, (3) has best regulation

3. Worst case for a load is with minimum resistance and minimum diode drop:

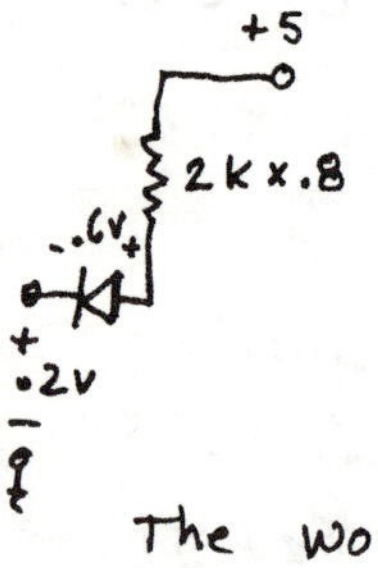

$$I_{LOAD}\Big|_{max} = \frac{5-.6-.2}{.8 \times 2K} = 2.63 \text{ mA per load}$$

The worst case of base drive is with maximum resistance and maximum diode / base-emitter drops

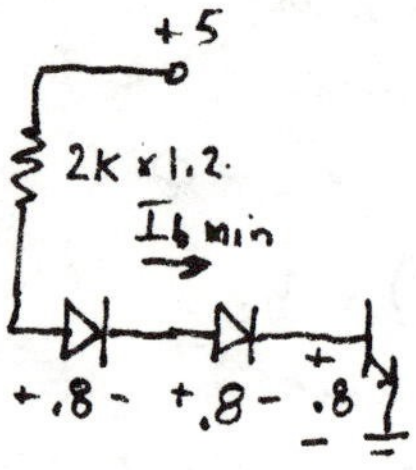

$$I_{b\,min} = \frac{5-.8-.8-.8}{2K \times 1.2} = 1.08 \text{ mA}$$

the worst case of total collector current is

$$I_c\Big|_{min} = 1.08\,\beta_{min} = 19.44 \text{ mA}$$

the maximum part of Ic that flows through the collector resistance is $I_s\Big|_{max} = \frac{5-.2}{.8 \times 2K} = 3$ mA

So the minimum SINKED current is

$$I_{sink}\Big|_{min} = 19.44 - 3 = 16.44 \text{ mA}$$

Then the maximum number of loads is

$$\frac{16.44 \text{ mA}}{2.63 \text{ mA}} = 6.25 \rightarrow 6 \text{ loads}$$

4. (a) As no current initially flows in the inductance, a current of $\frac{200}{2} = 100$ A must find a path through the lamp. In doing so it requires a 50 Volt lamp drop, so the initial current is $\frac{200-50}{2} = 75$ A. The lamp illuminates at $t=0$

(b) As long as the lamp is illuminated, a voltage of $50\text{V} = L\frac{di_L}{dt}$ exists across the inductance. this causes an inductance current:

$$i_L(t) = \frac{1}{L}\int 50\,dt = 50t + \text{constant.}$$

As $i_L(0) = 0$, the constant is zero

the capacitance current, as long as the lamp is illuminated, is

$$75e^{-t/RC} = 75e^{-25t}$$

the lamp will extinguish when its current reaches zero:

$$I_{lamp} = 75e^{-25t} - 50t \rightarrow 0$$

this occurs when $e^{-25t} = \frac{2}{3}t$

A few points are needed to find a starting point for the iterative numerical solution:

t	$\frac{1}{25}$	$\frac{2}{25}$	$\frac{3}{25}$
e^{-25t}	.368	.135	.05
$\frac{2}{3}t$	.027	.054	.08

take $t = \frac{3}{25} = .012$ as a starting point.

The formula $t_2 = \frac{3}{2}e^{-25t_1}$ does not converge (try it). So the alternative is used $t_2 = \frac{1}{25}\ln\frac{3}{2t_1}$

t_1	.12	.101	.108	.105	.106	.106
t_2	.101	.108	.105	.106	.106	.106

then the lamp extinguishes at $t = .106$ S

5 The UJT "fires" during the positive half-cycle of Vs. There after it begins to charge again, and may or may not "fire" again during the positive half-cycle. It does not matter, as the SCR is already turned "on". Each time the UJT has "fired" a transient involving $e^{-t/RC} = e^{-1000t}$ will occur. During the negative half cycle of Vs, which takes a time of $\frac{1}{120}$ sec

5. Continued

Any transient will have died out $\left[e^{-\frac{1000}{120}} = 2.4\times10^{-4}\right]$; so, when the positive half-cycle begins the capacitance voltage has reached the sinusoidal steady state:

$$V_C = \frac{V_S}{j\omega RC+1} = \frac{120\sqrt{2}}{1+j.377}$$

$$V_C = 159 \sin(377t - 20.6°)$$

By voltage division $V_E = 27$ volts. when V_C reaches this voltage, the UJT "fires":

$$27 = 159 \sin(377t_f - 20.6°)$$

so $$377t_f = \theta_f = 30.4°$$

6. R_S must be chosen so that the d.c. loadline intersects the negative resistance region:

$$R_S \geqq \frac{48}{10mA} = 4.8K \quad : \text{choose } R_S = 5K$$

Region 1: $\dfrac{V_D}{I_D} = \dfrac{28V}{1mA} = 28K\Omega$

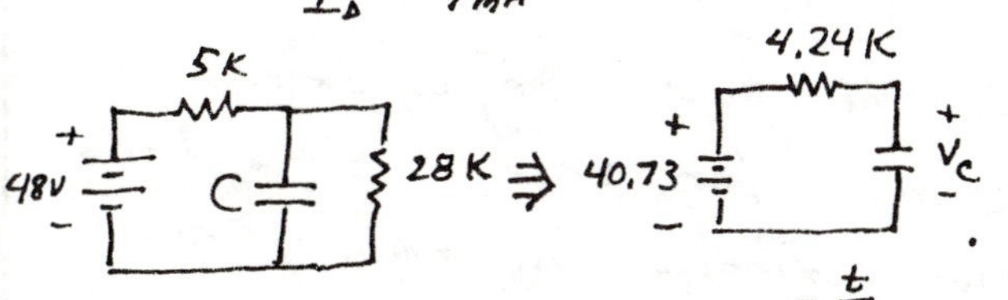

$$V_C = 40.73 + (9 - 40.73)e^{-\frac{t}{\tau_1}}$$

region 1 operation ends when $V_C = 28V$

$$28 = 40.73 - 31.73\,e^{-t_1/\tau_1}$$

$$t_1 = \tau_1 \ln\frac{31.73}{40.73-28} = .913\,\tau_1$$

Region 2: $\dfrac{V_D - 9}{I_D - .01} = \dfrac{28-9}{.2-.1}$

$$V_D = 100I_d + 8$$

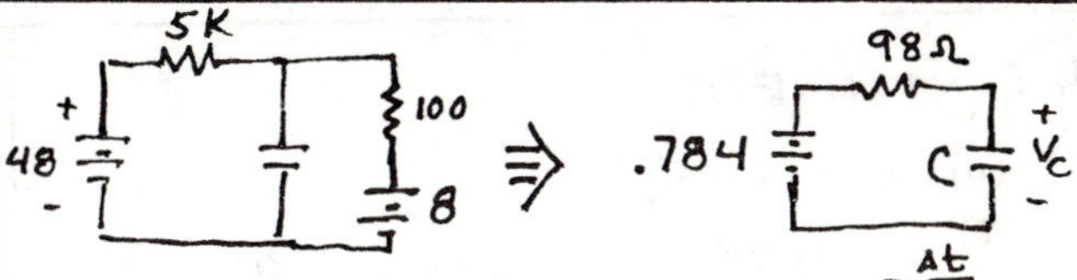

$$V_C = .784 + (28 - .784)e^{-\frac{\Delta t}{\tau_2}}$$

region 2 operation ends when V_C reaches 9 volts:

$$9 = .784 + 27.22\,e^{-\frac{\Delta t_2}{\tau_2}}$$

$$\Delta t_2 = \tau_2 \ln\frac{27.22}{9-.784} = 1.198\,\tau_2$$

$\tau_1 = 4240$ $\qquad \tau_2 = 98C$

$$T = \tau_1 + \tau_2 = 4338\,C = \frac{1}{400}$$

$$C = .576\,\mu F \rightarrow .58\,\mu F$$

$\tau_1 = 2.46$ mS , $\tau_2 = 57\,\mu S$

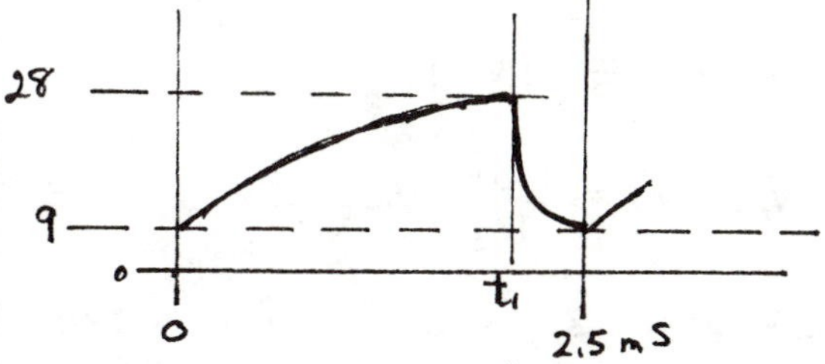

$t_1 \approx 2.44$mS $\qquad |\Delta t_2|$

$\Delta t_2 = .06$mS

TIMED

1. First consider the left-hand amplifier:

(a) with $V_{in} > 0$:

$V_{out}(1) = -\frac{R_1}{100K}V_{in}$, as the upper diode is conducting and behaves as a precision diode. $V_{out}(1)$ is taken between R_1 and R_2.

(b) with $V_{in} < 0$:

$V_{out}(1)$ is clamped to zero volts by the lower diode: $V_{out}(1) = 0$

NEXT consider the right-hand Amplifier:

1. CONTINUED

(a) with $V_{in} > 0$:

$$V_o = -\frac{100K}{R_3} V_{in} - \frac{100K}{R_2}\left(-\frac{R_1}{100K} V_{in}\right)$$
$$= |V_{in}|\left(\frac{R_1}{R_2} - \frac{100K}{R_3}\right)$$

(b) with $V_{in} < 0$

$$V_o = -\frac{100K}{R_3} V_{in} - \frac{100K}{R_2}(0)$$
$$= |V_{in}| \frac{100K}{R_3}$$

As this has been specified to be a full-wave rectifier, from (a) and (b) above:

$$\frac{R_1}{R_2} - \frac{100K}{R_3} = \frac{100K}{R_3}, \text{ or } \frac{R_1}{R_2} = \frac{200K}{R_3}$$

then $V_o = \frac{100K}{R_3} |V_{in}|$

or $V_o(AVG) = \frac{100K}{R_3} |V_{in}|(AVG)$

FROM chapter 3, p 3-3, for a full-wave rectified sinusoid:

$$V(AVG) = \frac{2}{\pi} V(peak) = \frac{2\sqrt{2}}{\pi} V(RMS)$$

then

$$V_o(AVG) = \frac{100K}{R_3} \frac{2\sqrt{2}}{\pi} V_{in}(RMS)$$

specification $V_o(AVG) = V_{in}(RMS)$ so $R_3 = \frac{200\sqrt{2}K}{\pi} = 90K\Omega$

as $\frac{R_1}{R_2} = \frac{200K}{R_3}$, set $R_1 = 200K\Omega$, $R_2 = R_3 = 90K\Omega$

The maximum output of either amplifier is 12 V. For the left-hand Amplifier

$$\frac{V_o}{V_{in}} = -\frac{R_1}{100K} = -2, \quad V_{in} \le \frac{12}{2} = 6$$

so $V_{in}(RMS)|_{max} = \frac{V_{in}(max)}{\sqrt{2}} = \frac{6}{\sqrt{2}} = 4.24$ V(RMS)

2. With at least one input low, the collector is high, and no current flows in the collector circuit. the current through the left resistance is found for an input voltage of .2 V and a diode drop of .6 volts to be

$$I_1 = \frac{4-.2-.6}{2K} = 1.6 \text{ mA}$$

So the power from the 4-volt source: $P_{high} = 1.6\text{ mA} \times 4\text{ V} = 6.4\text{ mW}$

With both inputs high, the input diodes are "off", and the left resistor has

$$I_1 = \frac{4-.6-.6-.6}{2K} = 1.1 \text{ mA}$$

The current from the 4 volt source through the right resistance is

$$I_2 = \frac{4-.2}{2K} = 1.9 \text{ mA}$$

then the power supplied by the 4 volt source is

$$P_{low} = 4(1.9+1.1)\text{mA} = 12\text{ mW}$$

So the power requirement is 12 mW for this stage. The power dissipation ability must include the sinked currents which would add 5 × .2V × 1.6 mA = 1.6 mW. So the power required for the stage is 12 mW and the power dissipation 13.6 mW

3. Follow example 9.13, p 9-21

$\omega RC = 377 \times 10^5 \times 10^{-8} = .377$

$V_c(ss) = \dfrac{V_s}{1+j.377} = \dfrac{220\sqrt{2}\angle 0}{1+j.377}$

$V_c(ss) = 291.1 \sin(377t - .3605)$ ← radians

as $V_{hold} \approx 0$, the diac will begin each half-cycle at 0.0 volts

$V_c(t) = Ae^{-t/RC} + V_c(s.s.)$

$V_c(0) = 0 = A + 291.1 \sin(-.3605)$

$A = 102.68$

Iteration formula

$4 = 102.68 e^{-10^3 t_f} + 291.1 \sin(377 t_f - .3605)$

$t_{f2} = \dfrac{1}{377}\left[.3605 + \sin^{-1} \dfrac{4 - 102.68 e^{-10^3 t_{f1}}}{291.1}\right]$

Initial guess

$t_{f1} = \dfrac{1}{377}[.3605] = .00096$

$t_{f2} = .00063$	$t_{f6} = .00034$
$t_{f3} = .00049$	$t_{f7} = .00032$
$t_{f4} = .00042$	$t_{f8} = .00031$
$t_{f5} = .00037$	$t_{f9} = .00030$

$t_{f10} = .00029$

$t_{f11} = .00029$

$t_{f12} = 2.86 \times 10^{-4}$

$t_{f13} = 2.83 \times 10^{-4} \rightarrow 2.7 \times 10^{-4}$ S

try $t_{f1} = .000270$, $t_{f2} = .000271$

try $t_{f1} = .000275$, $t_{f2} = .0002745$

$T_{half\text{-}cycle} = \dfrac{1}{120} = 8.33 \times 10^{-3}$

$\dfrac{t_f}{T_{h\text{-}c}} = .032$, so the circuit fires almost immediately at the beginning of each half-cycle

The conduction angle is 180° − firing angle $= 180° - 377 \times 2.7 \times 10^{-4} \times \dfrac{180}{\pi}$

$= 180° - 5.8° = 174.2°$ (a)

(b) $I_{gt} \times 50\,\mu S = Q_c = 4V \times 10^{-8} F = 4 \times 10^{-8}$ Coul

$I_{gt} = \dfrac{4 \times 10^{-8}}{50 \times 10^{-6}} = 0.8$ mA

(c)

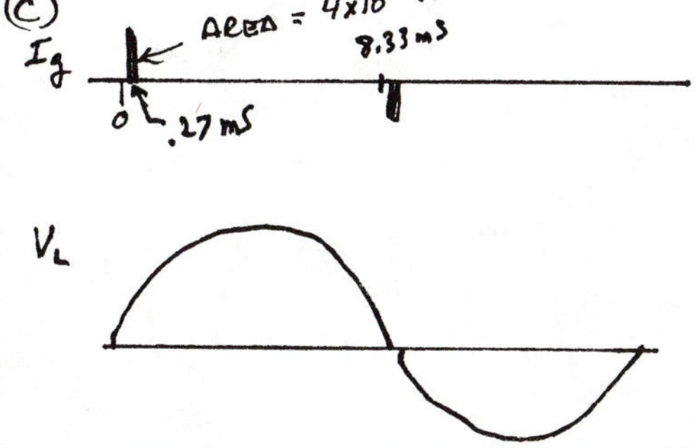

4. See fig. 9.35

Region A: begins with $I_d = I_L = .1$ mA at $t = 0$ and ends when $I_d = .4$ mA at $t = T_A$. from the graph $\dfrac{V_d}{I_d} = \dfrac{.1}{.0004} = 250$

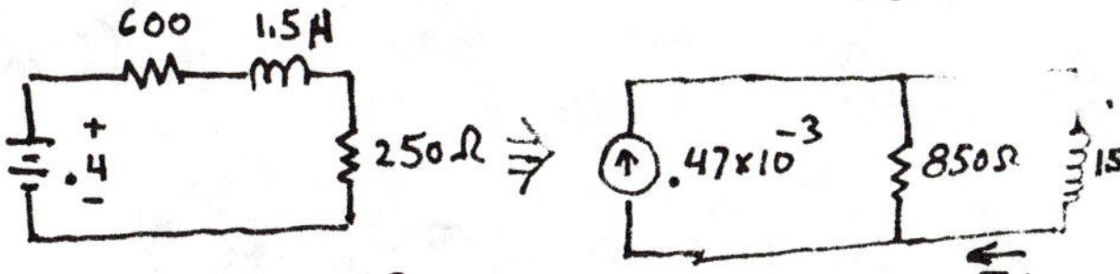

$\tau = \dfrac{1.5}{850} = \dfrac{3}{1700} = 1.765$ mS

$i_D = .47 \times 10^{-3} + (.1 \times 10^{-3} - .47 \times 10^{-3}) e^{-1700t/3}$

at $t = T_A$

$.4 \times 10^{-3} = \dfrac{.4}{850} + \left(10^{-4} - \dfrac{.4}{850}\right) e^{-\frac{1700}{3} T_A}$

$T_A = \dfrac{3}{1700} \ln \dfrac{10^{-4} - \frac{.4}{850}}{.4 \times 10^{-3} - \frac{.4}{850}} = 2.93$ mS

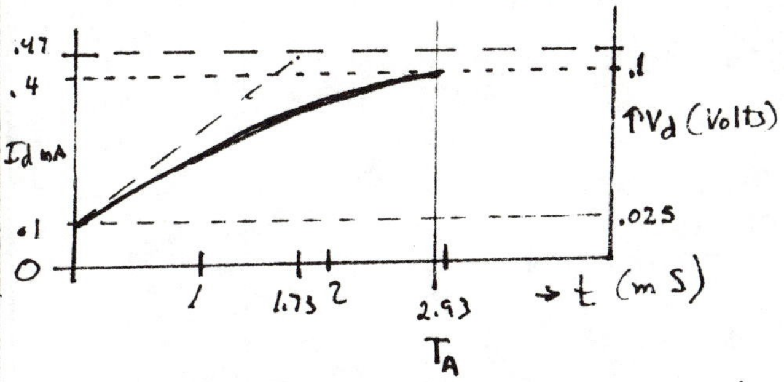

REGION C: begins with $I_d = .4$ mA at $t = T_A$, and ends with $I_d = .1$ mA at T.

Equivalent circuit on next page.

4 CONTINUED

from graph on page 9-33

$$\frac{V_d - .4}{I_d - 10^{-4}} = \frac{.8 - .4}{(4-1)\times 10^{-4}} \therefore V_d = \frac{.8}{3} + \frac{4000}{3} I_d$$

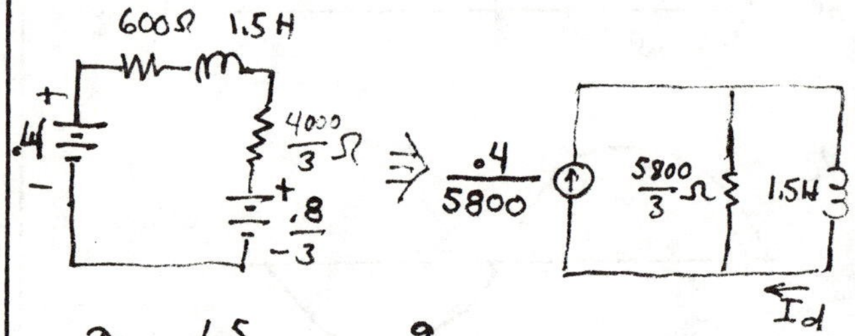

$$\tau = \frac{1.5}{\frac{5800}{3}} = \frac{9}{11600} = .78 \text{ mS}$$

$$i_d(t) = \frac{.4}{5800} + \left[.4\times10^{-3} - \frac{.4}{5800}\right] e^{-\frac{11600}{9}(t - T_A)}$$

$$= .069\times10^{-3} + (.331\times10^{-3}) e^{-1289(t-T_A)}$$

$$i_d(T) = 10^{-4} = .069\times10^{-3} + (.331\times10^{-3}) e^{-1289(T-T_A)}$$

$T - T_A = 1.84$ mS , $T = 1.84 + 2.93$
$= 4.77$ mS (a)

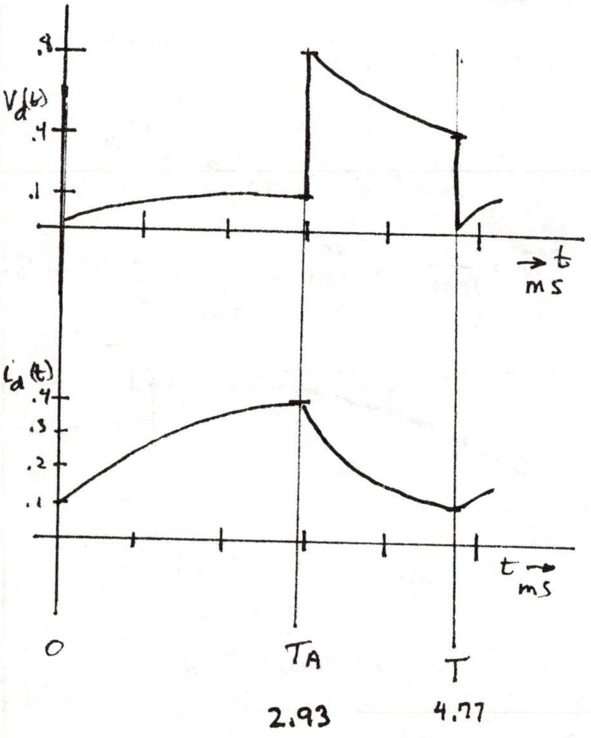

5.

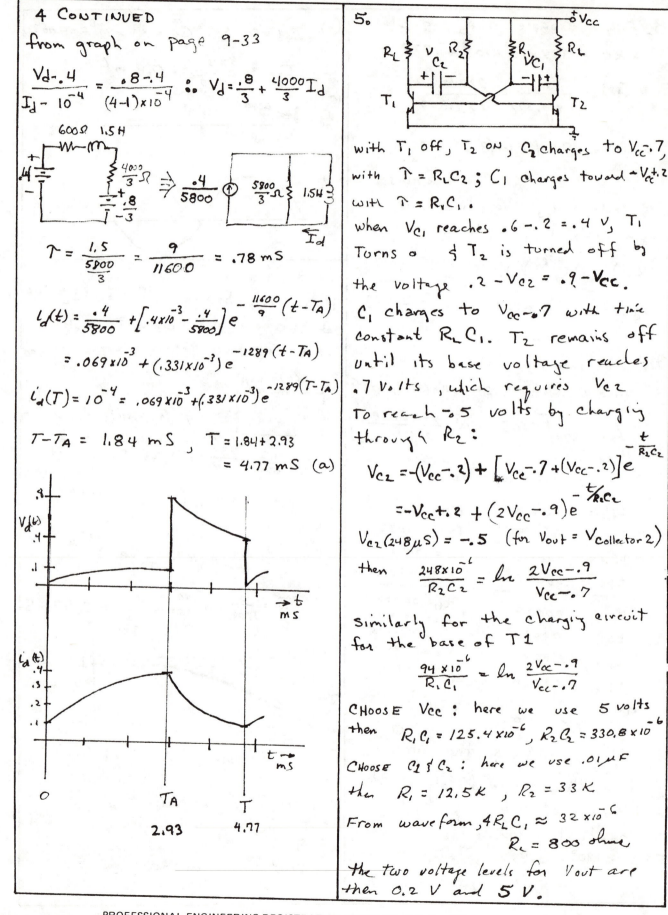

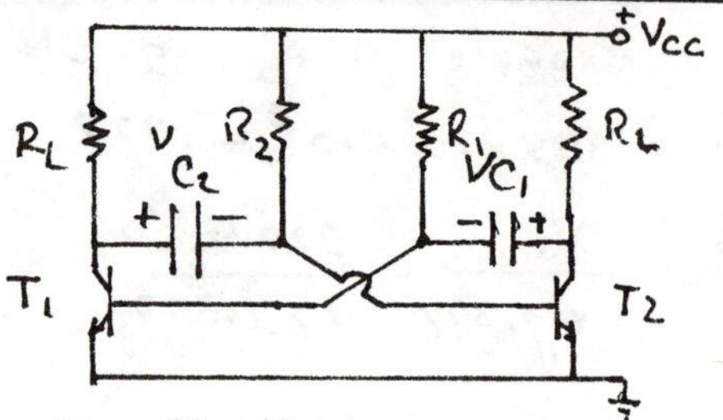

with T_1 off, T_2 ON, C_2 charges to $V_{cc} - .7$, with $\tau = R_L C_2$; C_1 charges toward $-V_{cc} + .2$ with $\tau = R_1 C_1$.

when V_{C_1} reaches $.6 - .2 = .4$ V, T_1 Turns o & T_2 is turned off by the voltage $.2 - V_{C2} = .9 - V_{cc}$.

C_1 charges to $V_{cc} - .7$ with time constant $R_L C_1$. T_2 remains off until its base voltage reaches .7 volts, which requires V_{C2} To reach $-.5$ volts by charging through R_2:

$$V_{C2} = -(V_{cc} - .2) + \left[V_{cc} - .7 + (V_{cc} - .2)\right] e^{-\frac{t}{R_2 C_2}}$$

$$= -V_{cc} + .2 + (2V_{cc} - .9) e^{-t/R_2C_2}$$

$V_{C2}(248\mu S) = -.5$ (for $V_{out} = V_{collector\,2}$)

then $\frac{248\times10^{-6}}{R_2 C_2} = \ln \frac{2V_{cc} - .9}{V_{cc} - .7}$

similarly for the charging circuit for the base of T1

$$\frac{94\times10^{-6}}{R_1 C_1} = \ln \frac{2V_{cc} - .9}{V_{cc} - .7}$$

CHOOSE V_{cc} : here we use 5 volts

then $R_1 C_1 = 125.4\times10^{-6}$, $R_2 C_2 = 330.8\times10^{-6}$

CHOOSE C_1 & C_2 : here we use .01μF

then $R_1 = 12.5K$, $R_2 = 33K$

From waveform, $4 R_L C_1 \approx 32\times10^{-6}$

$R_L = 800$ ohms

the two voltage levels for V_{out} are then 0.2 V and 5 V.

WARMUPS

1. $D = \overline{(A\cdot B)\cdot(B+C)}$

A B C	A·B	B+C	$\bar{D}$	D
0 0 0	0	0	0	1
0 0 1	0	1	0	1
0 1 0	0	1	0	1
0 1 1	0	1	0	1
1 0 0	0	0	0	1
1 0 1	0	1	0	1
1 1 0	1	1	1	0
1 1 1	1	1	1	0

2. $D = \bar{A}\cdot\bar{B}\cdot\bar{C} + \bar{A}\cdot\bar{B}\cdot C + \bar{A}\cdot B + \bar{C}$
$+ \bar{A}\cdot B\cdot C + A\cdot\bar{B}\cdot\bar{C} + A\cdot\bar{B}\cdot C$

$\bar{D} = A\cdot B\cdot\bar{C} + A\cdot B\cdot C$

$D = (\bar{A}+\bar{B}+C)\cdot(\bar{A}+\bar{B}+\bar{C})$

3. $D = \bar{A} + \bar{B}$ $\qquad \bar{D} = A\cdot B$ demorgan

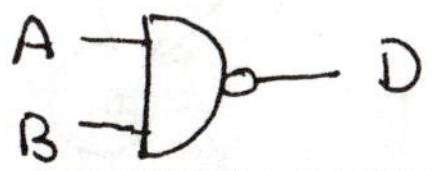

4.

(D)

BC \ A	0	1
00	0	1
01	0	1
11	0	0
10	X	X

(E)

BC \ A	0	1
00	1	1
01	X	1
11	0	1
10	1	X

(F)

BC \ A	0	1
00	1	X
01	1	X
11	0	X
10	0	X

5. $D = A\cdot\bar{B}$

$E = \bar{B}+\bar{C} = \overline{\overline{\bar{B}+\bar{C}}} = \overline{B\cdot C}$ demorgen

$= A + \bar{C}$

$F = \bar{B}$

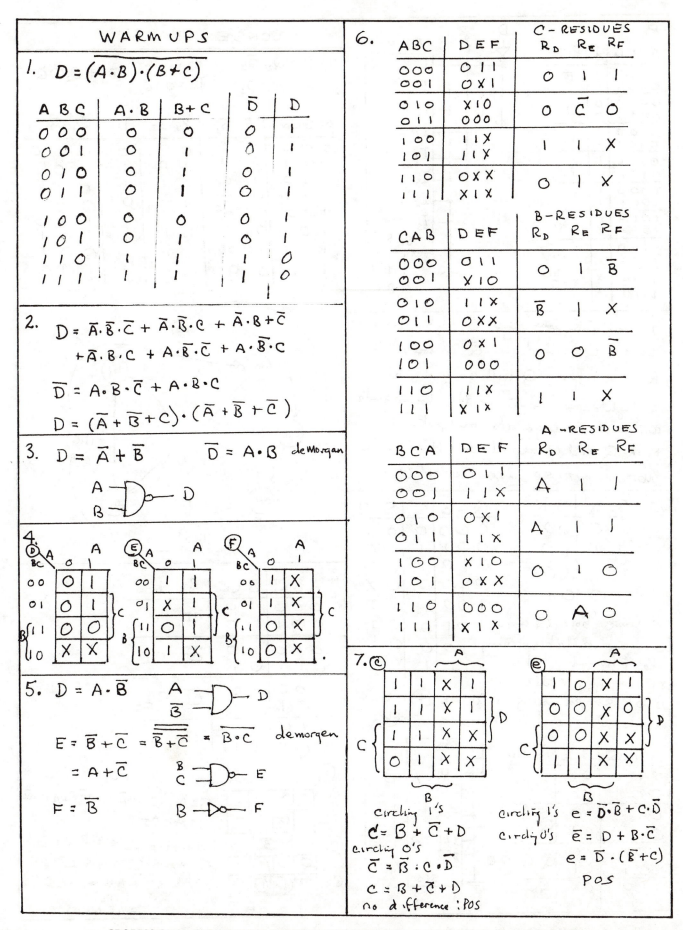

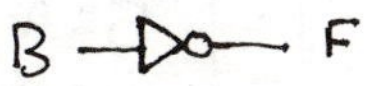

6.

ABC	DEF	C-RESIDUES R_D	R_E	R_F
000 001	011 0X1	0	1	1
010 011	X10 000	0	$\bar{C}$	0
100 101	11X 11X	1	1	X
110 111	0XX X1X	0	1	X

CAB	DEF	B-RESIDUES R_D	R_E	R_F
000 001	011 X10	0	1	$\bar{B}$
010 011	11X 0XX	$\bar{B}$	1	X
100 101	0X1 000	0	0	$\bar{B}$
110 111	11X X1X	1	1	X

BCA	DEF	A-RESIDUES R_D	R_E	R_F
000 001	011 11X	A	1	1
010 011	0X1 11X	A	1	1
100 101	X10 0XX	0	1	0
110 111	000 X1X	0	A	0

7. (c)

1	1	X	1
1	1	X	1
1	1	X	X
0	1	X	X

circling 1's
$c = B + \bar{C} + D$
circling 0's
$\bar{c} = \bar{B} : C\cdot\bar{D}$
$c = B + \bar{C} + D$
no difference : POS

(e)

1	0	X	1
0	0	X	0
0	0	X	X
1	1	X	X

circling 1's $e = \bar{D}\cdot\bar{B} + C\cdot\bar{D}$
circling 0's $\bar{e} = D + B\cdot\bar{C}$
$e = \bar{D}\cdot(\bar{B}+C)$
POS

8.

ABCD	f	RES
0000	1	$\bar{D}$
0001	0	
0010	0	0
0011	0	
0100	1	1
0101	1	
0110	1	$\bar{D}$
0111	0	
1000	1	1
1001	1	
1010	X	X→0
1011	X	
1100	X	X→0
1101	X	
1110	X	X→0
1111	X	

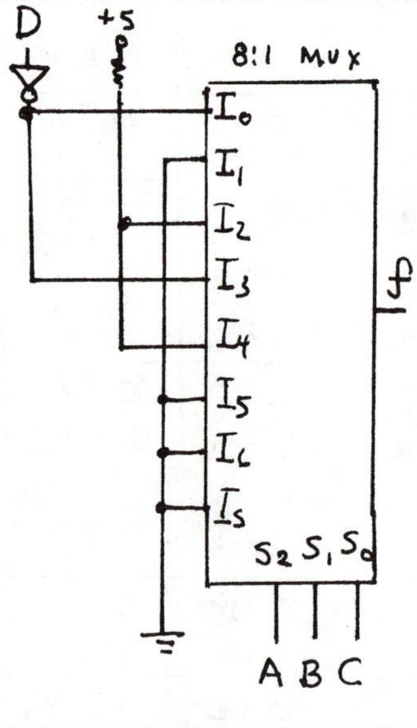

9. With R low and S high for some time, Q is high and $\bar{Q}$ is low. When R goes high, Q will go low, but a high-low input on the lower gate will not change its output. Therefore both Q and $\bar{Q}$ are low; this is a "NOT PERMITTED" input condition.

10.

Q	Q^+	S	R
0	0	0	X
0	1	1	0
1	1	X	0
1	0	0	1

QAB	Q^+	S	R
000	0	0	X
001	1	1	0
010	1	1	0
011	0	0	X
100	1	X	0
101	0	0	1
110	0	0	1
111	1	X	0

because both Q and $\bar{Q}$ are available from the flip-flop, the minimum circuit is obtained using a DUAL 4:1 MUX with Q-residues. Rearrange table:

ABQ	S R	R_S	R_R
000	0 X	0	0
001	X 0		
010	1 0	$\bar{Q}$	Q
011	0 1		
100	1 0	$\bar{Q}$	Q
101	0 1		
110	0 X	0	0
111	X 0		

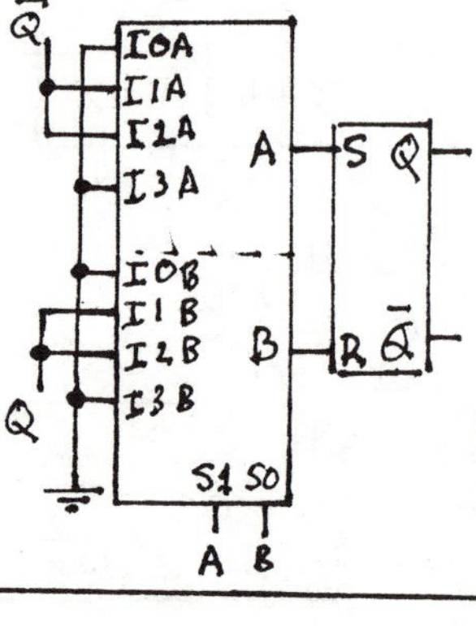

CONCENTRATES

1. $R_1 = Q_2 \cdot \bar{Q}_3$ $K2 = Q_1 + Q_3$ $D_3 = Q_2 \oplus Q_3$

$S_1 = \bar{Q}_2 \cdot Q_3$ $J2 = Q_1 + \bar{Q}_3$

$Q_1Q_2Q_3$	R_1	S_1	K2	J2	D_3	Q_1^+	Q_2^+	Q_3^+
000	0	0	0	1	0	0	1	0
001	0	1	1	0	1	1	0	1
010	1	0	0	1	1	0	1	1
011	0	0	1	0	0	0	1	1
100	0	0	1	1	0	0	0	0
101	0	1	1	1	1	1	1	1
110	1	0	1	1	1	0	0	1
111	0	0	1	1	0	1	0	0

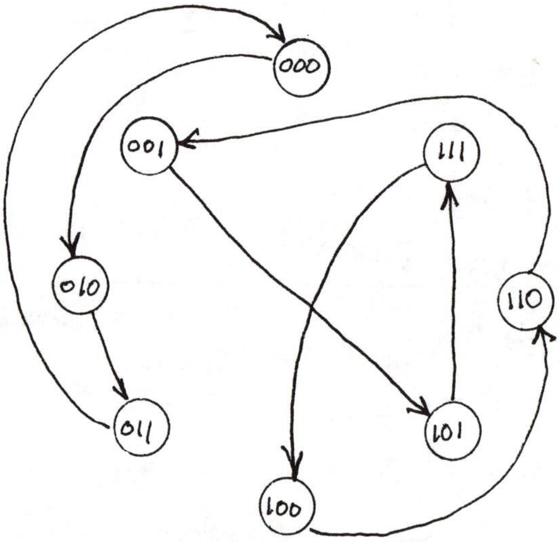

2.

ABCI	S	CO	R_S	R_{CO}
000	0	0	CI	0
001	1	0		
010	1	0	$\overline{CI}$	CI
011	0	1		
100	1	0	$\overline{CI}$	CI
101	0	1		
110	0	1	CI	1
111	1	1		

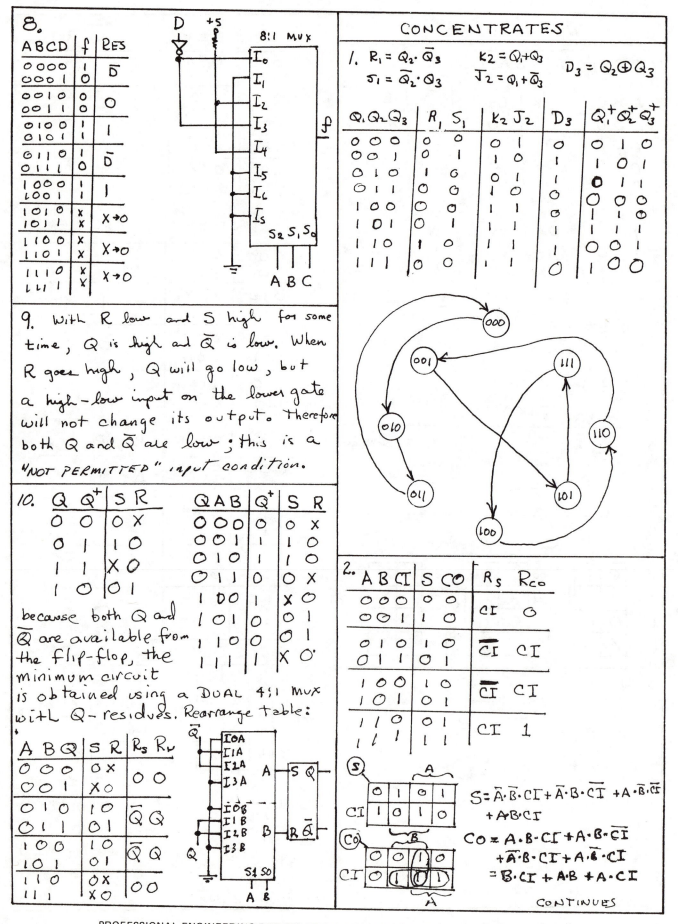

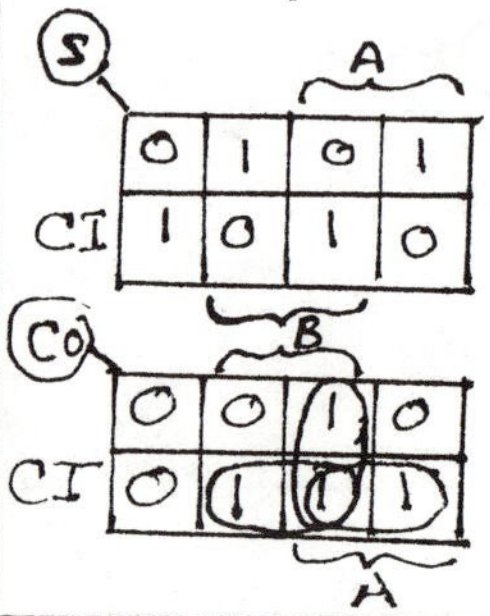

$S = \bar{A} \cdot \bar{B} \cdot CI + \bar{A} \cdot B \cdot \overline{CI} + A \cdot \bar{B} \cdot \overline{CI} + A \cdot B \cdot CI$

$CO = A \cdot B \cdot CI + A \cdot B \cdot \overline{CI} + \bar{A} \cdot B \cdot CI + A \cdot \bar{B} \cdot CI$
$= B \cdot CI + A \cdot B + A \cdot CI$

CONTINUES

2. CONTINUED

(a)

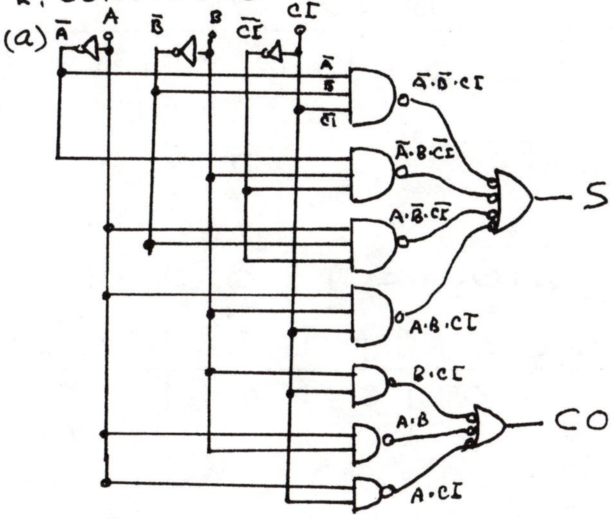

(b)

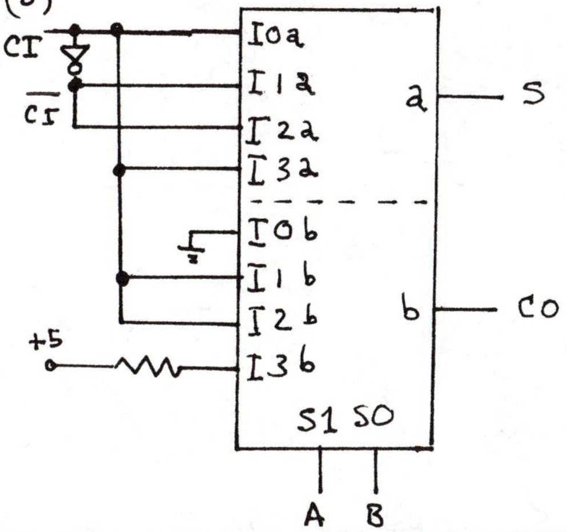

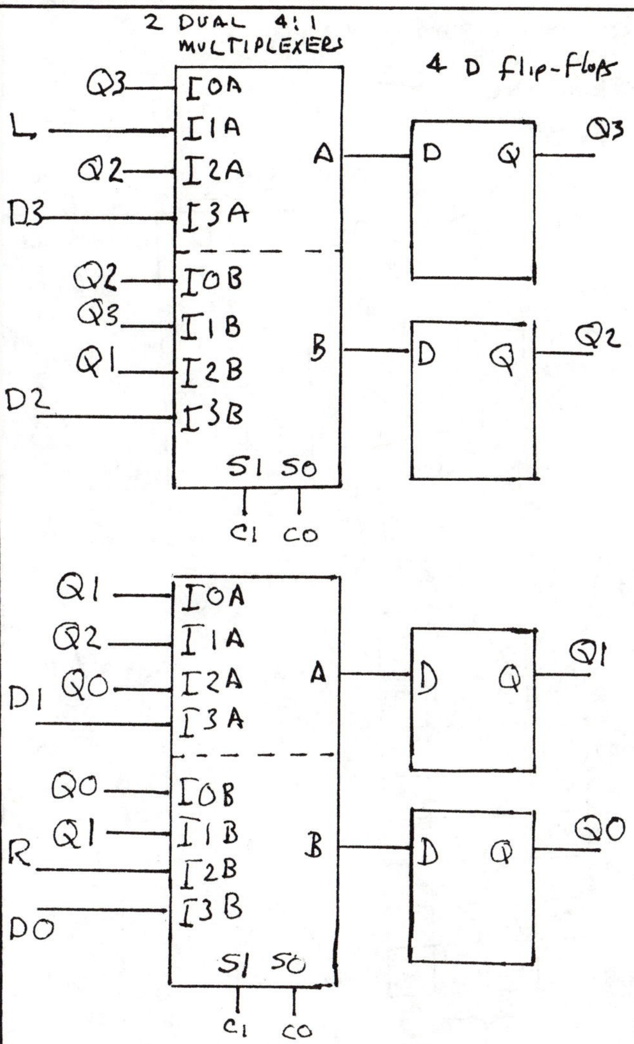

3. This is done using a basic cell:

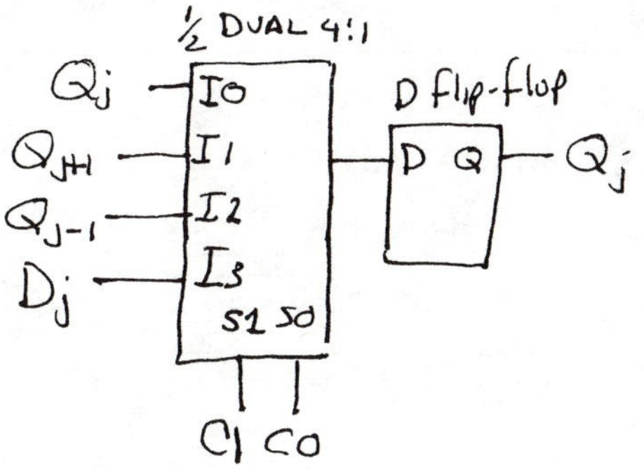

where

j	Q_j	Q_{j+1}	Q_{j-1}	D_j
3	Q_3	L	Q_2	D_3
2	Q_2	Q_3	Q_1	D_2
1	Q_1	Q_2	Q_0	D_1
0	Q_0	Q_1	R	D_0

4. $K_A = \overline{Q_B} = \bar{B}$ $J_A = \overline{\bar{B}+C} = B\cdot\bar{C}$

$K_B = A$ $J_B = C$

$K_C = B$ $J_C = \overline{A+B} = \bar{A}\cdot\bar{B}$

A B C	K_A	J_A	K_B	J_B	K_C	J_C	A^+	B^+	C^+
0 0 0	1	0	0	0	0	1	0	0	1
0 0 1	1	0	0	1	0	1	0	1	1
0 1 0	0	1	0	0	1	0	1	1	0
0 1 1	0	0	0	1	1	0	0	1	0
1 0 0	1	0	1	0	0	0	0	0	0
1 0 1	1	0	1	1	0	0	0	1	1
1 1 0	0	1	1	0	1	0	1	0	0
1 1 1	0	0	1	1	1	0	1	0	0

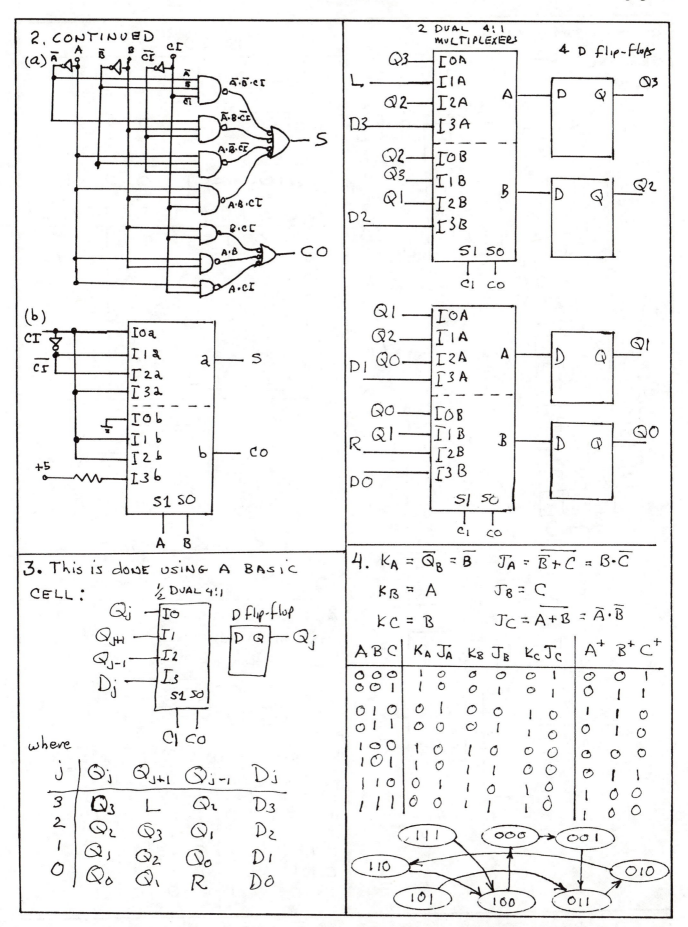

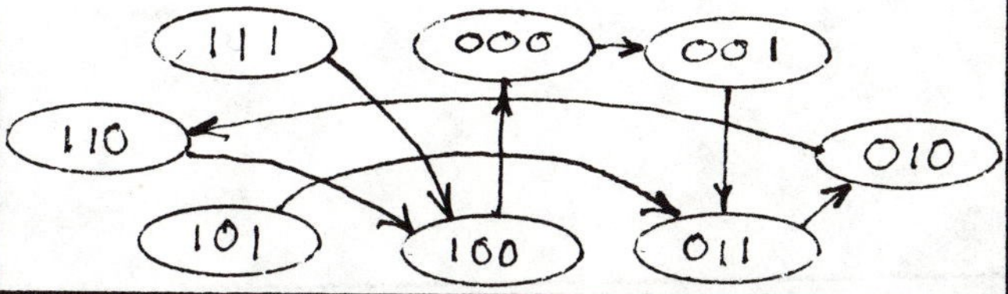

5. from timing diagram → SEQUENCE

A B C
0 0 0
1 0 0
1 1 0
0 1 1
0 0 1
1 0 1
1 1 1
0 0 0

NEXT STATE TABLE

ABC	$A^+B^+C^+$	T_A	T_B	T_C
000	100	1	0	0
001	101	1	0	0
010	---	X	X	X
011	001	0	1	0
100	110	0	1	0
101	111	0	1	0
110	011	1	0	1
111	000	1	1	1

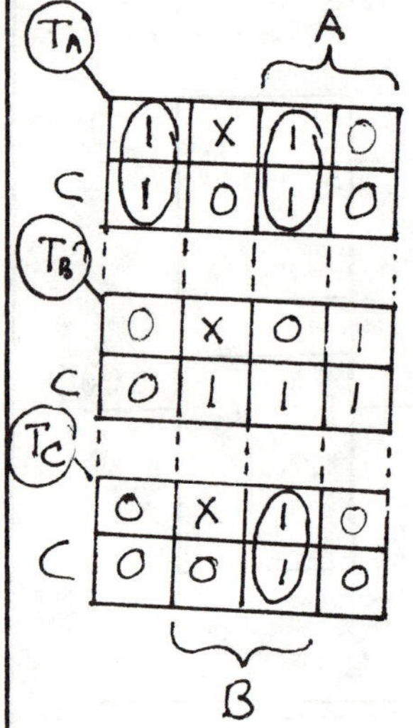

$T_A = A\cdot B + \bar{A}\cdot\bar{B} = \overline{A\oplus B}$

$T_B = B\cdot C + A\cdot B$

$T_C = A\cdot B$

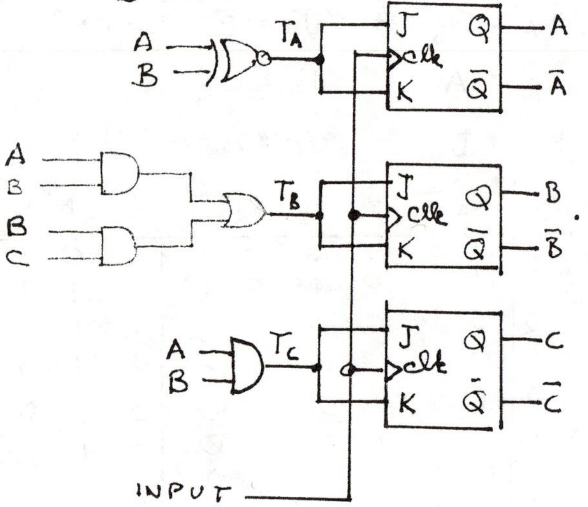

TIMED

1. $\overline{\overline{Q_1\cdot Q_2}\cdot Q_3} = K = \bar{Q}_3 + Q_2Q_1$

$J = Q_2$

$\overline{\bar{Q}_1 + Q_2} = D = \bar{Q}_2\cdot Q_1$

$S = \bar{Q}_2$

$\overline{\bar{Q}_3 + \bar{Q}_2} = R = Q_3\cdot Q_2$

$Q_3 Q_2 Q_1$	K	J	D	R	S	Q_3^+	Q_2^+	Q_1^+
0 0 0	1	0	0	0	1	0	0	1
0 0 1	1	0	1	0	1	0	1	1
0 1 0	1	1	0	0	0	1	0	0
0 1 1	1	1	0	0	0	1	0	1
1 0 0	0	0	0	0	1	1	0	1
1 0 1	0	0	1	0	1	1	1	1
1 1 0	0	1	0	1	0	1	0	0
1 1 1	1	1	0	1	0	0	0	0

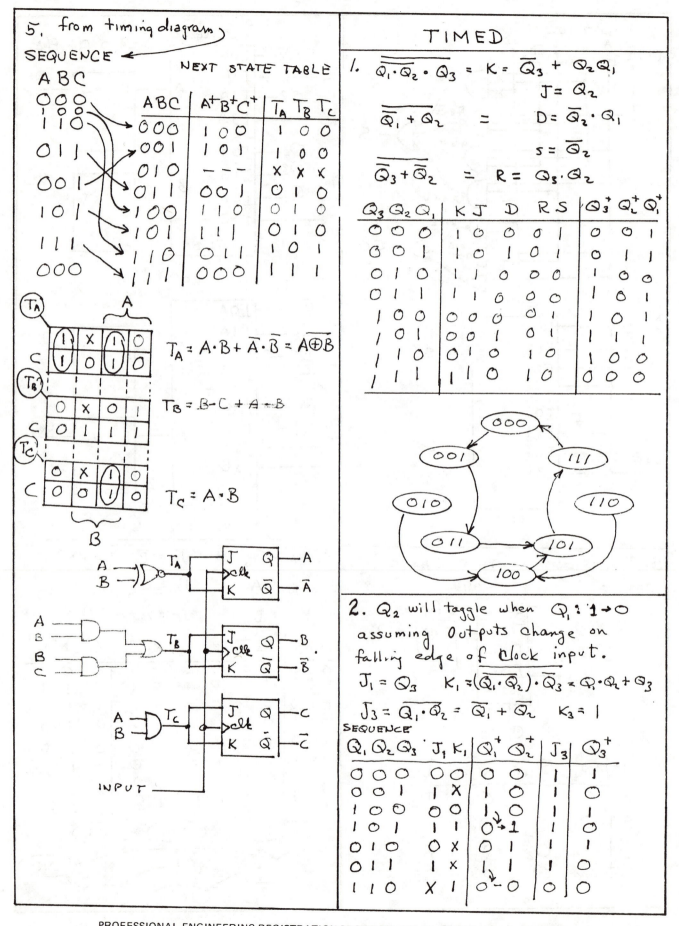

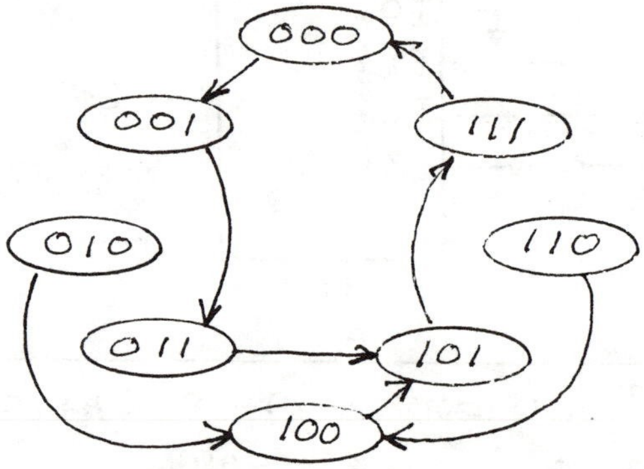

2. Q_2 will toggle when Q_1: 1→0 assuming outputs change on falling edge of clock input.

$J_1 = Q_3$ $K_1 = \overline{(\overline{Q_1\cdot Q_2})\cdot \bar{Q}_3} = Q_1\cdot Q_2 + Q_3$

$J_3 = \overline{Q_1\cdot Q_2} = \bar{Q}_1 + \bar{Q}_2$ $K_3 = 1$

SEQUENCE

$Q_1 Q_2 Q_3$	J_1	K_1	Q_1^+	Q_2^+	J_3	Q_3^+
0 0 0	0	0	0	0	1	1
0 0 1	1	X	1	0	1	0
1 0 0	0	0	1	0	1	1
1 0 1	1	1	0	1	1	0
0 1 0	0	X	0	1	1	1
0 1 1	1	X	1	1	1	0
1 1 0	X	1	0	0	0	0

2. continued

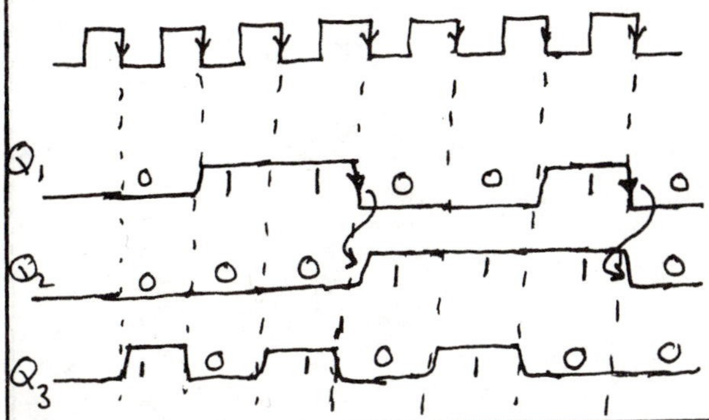

3. Use A as least significant variable in Veitch diagram:

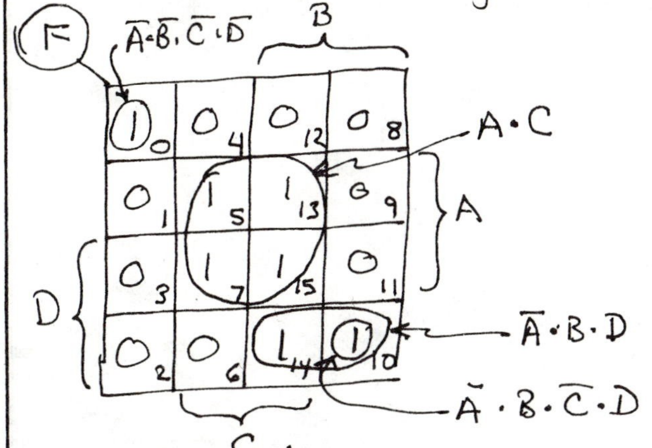

m	BCD	A	F	R	I
0	000	0	1	$\bar{A}$	0
1		1	0		
2	001	0	0	0	1
3		1	0		
4	010	0	0	A	2
5		1	1		
6	011	0	0	A	3
7		1	1		
8	100	0	0	0	4
9		1	0		
10	101	0	1	$\bar{A}$	5
11		1	0		
12	110	0	0	A	6
13		1	1		
14	111	0	1	1	7
15		1	1		

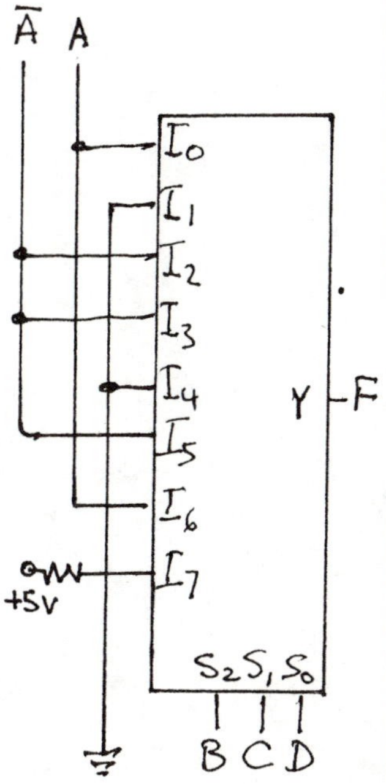

4.

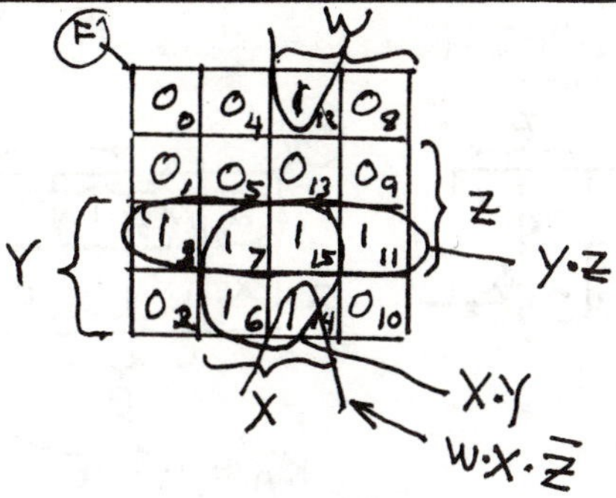

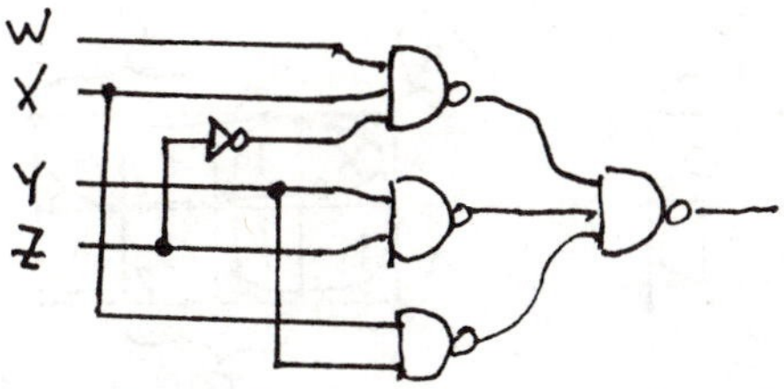

5

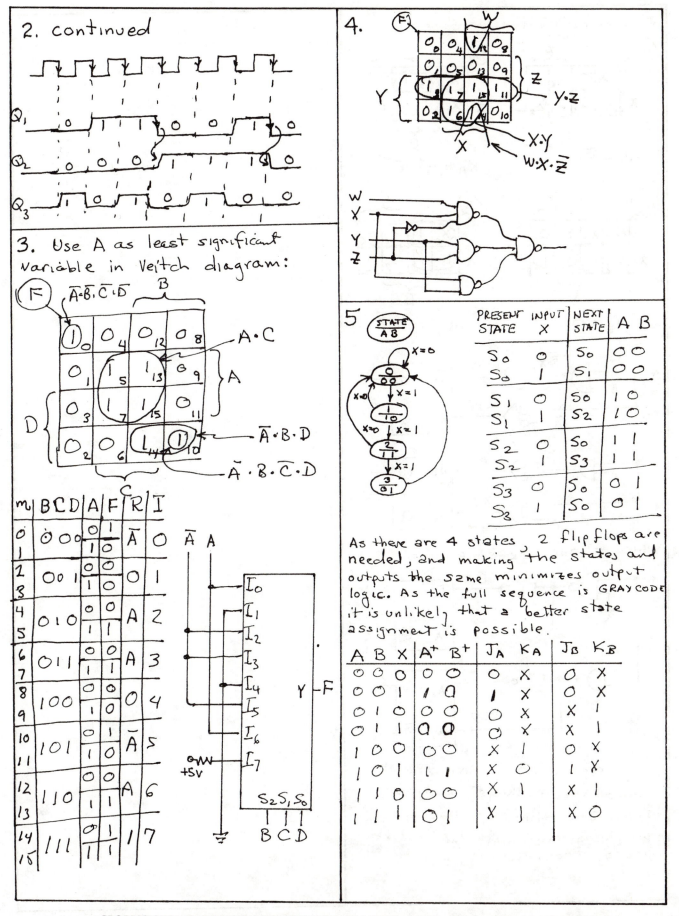

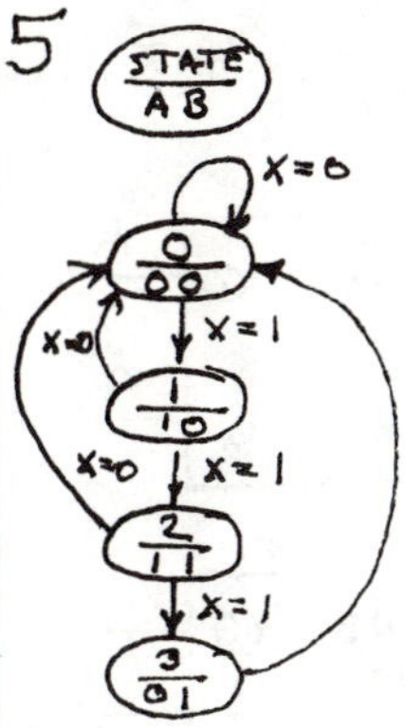

PRESENT STATE	INPUT X	NEXT STATE	A B
S_0	0	S_0	0 0
S_0	1	S_1	0 0
S_1	0	S_0	1 0
S_1	1	S_2	1 0
S_2	0	S_0	1 1
S_2	1	S_3	1 1
S_3	0	S_0	0 1
S_3	1	S_0	0 1

As there are 4 states, 2 flip flops are needed, and making the states and outputs the same minimizes output logic. As the full sequence is GRAY CODE it is unlikely that a better state assignment is possible.

A	B	X	A^+	B^+	J_A	K_A	J_B	K_B
0	0	0	0	0	0	X	0	X
0	0	1	1	0	1	X	0	X
0	1	0	0	0	0	X	X	1
0	1	1	0	0	0	X	X	1
1	0	0	0	0	X	1	0	X
1	0	1	1	1	X	0	1	X
1	1	0	0	0	X	1	X	1
1	1	1	0	1	X	1	X	0

5. Continued

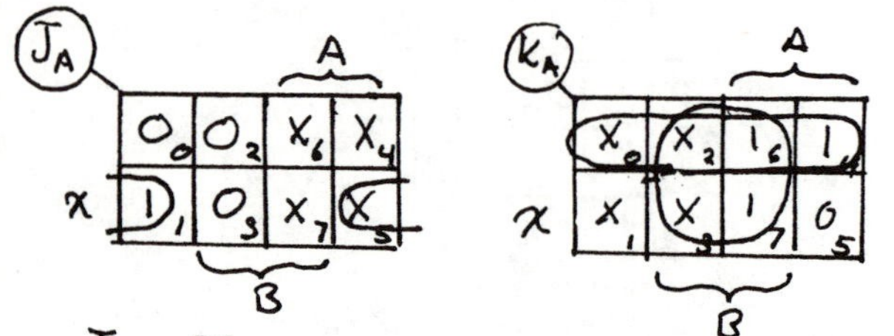

$J_A = \overline{B} \cdot x$

$K_A = B + \overline{x}$

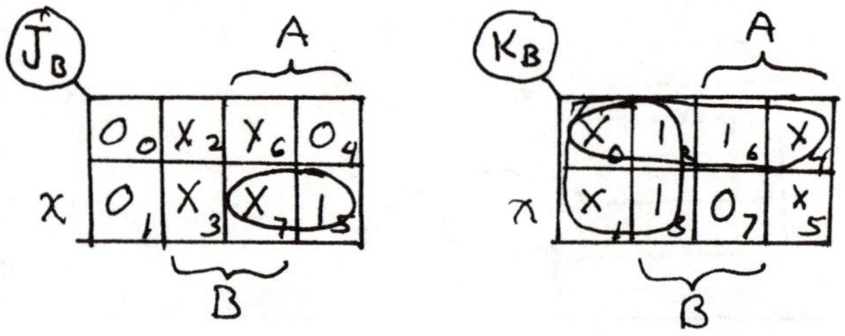

$J_B = A \cdot x$

$K_B = \overline{A} + \overline{x}$

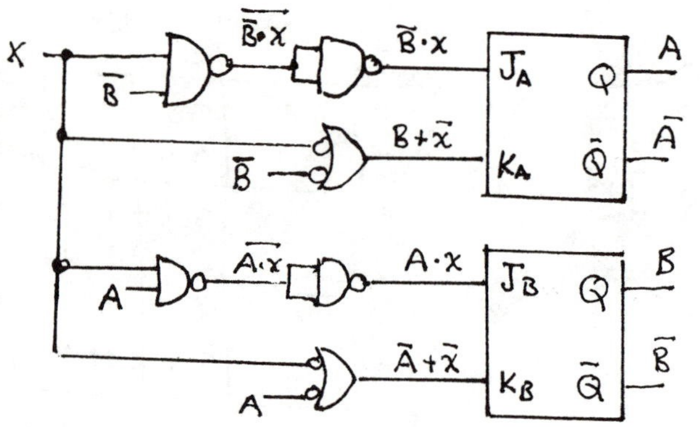

WHICH IS FURTHER SIMPLIFED

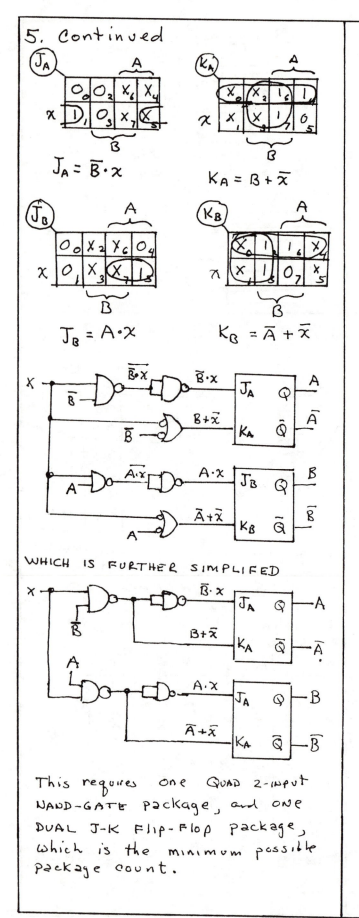

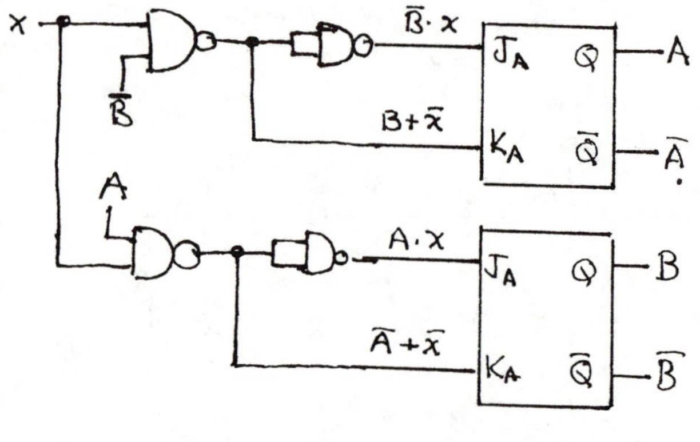

This requires one QUAD 2-INPUT NAND-GATE package, and ONE DUAL J-K Flip-Flop package, which is the minimum possible package count.

WARMUPS

1.

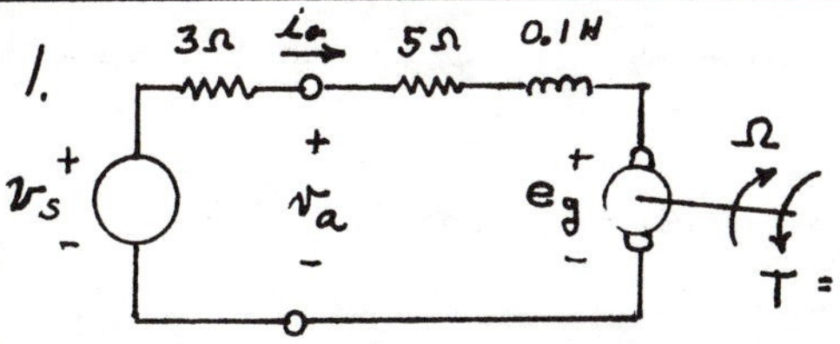

$T = D\Omega + J\frac{d\Omega}{dt}$

$v_a = 5 i_a + 0.1\frac{di_a}{dt} + e_g$

steady state ($\frac{di_a}{dt} = 0$) given:

$v_a = 24V$, $i_a = 1.6A$ then $e_g = 16$ V

given: $\Omega = 160$ rad/S

$e_g = K_V \Omega$ then $K_V = \frac{16}{160} = 0.1$

In MKS units $K_T = K_V$: $T = K_V I_a$

$T = D\Omega + J\frac{d\Omega}{dt}$ $T_{s.s.} = D\Omega$ (as $\frac{d\Omega}{dt} = 0$)

$D = \frac{K_V I_a}{\Omega} = \frac{0.1 \times 1.6}{160} = 10^{-3} \frac{N\text{-}m}{rad/S}$

$\frac{T}{D} = \frac{J}{D}\frac{d\Omega}{dt} + \Omega$ $\therefore \tau_{mech} = \frac{J}{D} = 1.85$ S (given)

$J = 1.85 \times 10^{-3}$

EQUATIONS (transformed)

$V_s = (8 + .1s) I_A + E_g$; $E_g = 0.1\Omega = .1s\Theta$

$T = 10^{-3}(1 + 1.85s)\Omega = 10^{-3}s(1 + 1.85s)\Theta$

$T = .1 I_A$

SOLVING: $\frac{\Theta}{s} = \frac{5.56}{s\left(\frac{s}{1.26}+1\right)\left(\frac{s}{79.3}+1\right)} = G_m(s)$

(a)

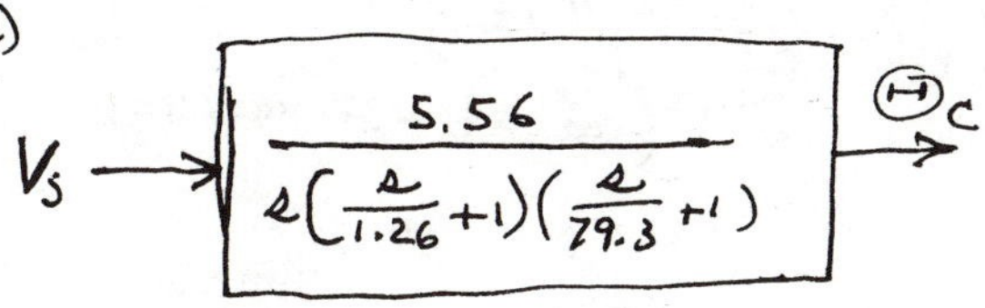

From fig. 11.2

$\frac{V_e}{R_f} + \frac{V_{tach}}{R_\Omega} + \frac{V_\Theta}{R_\Theta} + \frac{V_{in}}{R_{in}} = 0$

$V_{tach} = .05 s\Theta$, $V_\Theta = \frac{10}{\pi}\Theta_c$, $V_{in} = -\frac{10}{\pi}\Theta_{in}$

$V_S = V_e = \frac{10}{\pi}\left[\Theta_{in} - \Theta_c(1 + k_T s)\right]$

where $k_T = \frac{50\pi}{R_\Omega}$

$R(s) = \Theta_{in}$, $C(s) = \Theta_c$, $E(s) = \frac{\pi V_s}{10}$

(b) $E(s) = R(s) - C(s)(1 + k_T s)$

$C(s) = \frac{10}{\pi} A_p G_m(s) E(s)$

(c)

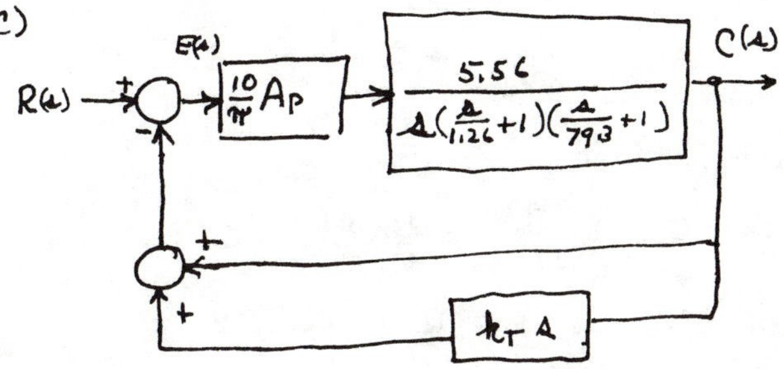

2. $\frac{10(s+2)}{s(s+20)} = \frac{\frac{s}{2}+1}{s\left(\frac{s}{20}+1\right)} = G$

$20\log|G| = 20\log\left|\frac{s}{2}+1\right| - 20\log|s| - 20\log\left|\frac{s}{20}+1\right|$

$20\log|s| = 0$ at $\omega = 1$

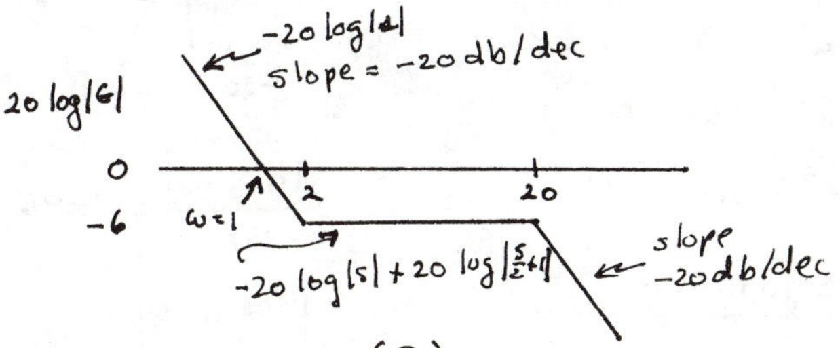

(a)

(b) $r(t) = u(t)$; $R(s) = \frac{1}{s}$

$e_{ss}(t) = \lim_{s\to 0} \frac{sR(s)}{1+G} = \lim_{s\to 0}\frac{s\frac{1}{s}}{1+\frac{1}{s}} = 0$

(c) $r(t) = t\,u(t)$: $R(s) = \frac{1}{s^2}$

$e_{ss}(t) = \lim_{s\to 0}\frac{s\frac{1}{s^2}}{1+\frac{1}{s}} = \lim_{s\to 0}\frac{1}{s+1} = 1$

3. $G = \frac{50s}{s^2+10s+400} = \frac{1}{8}\frac{s}{\left(\frac{s}{20}\right)^2 + 2\times.25\left(\frac{s}{20}\right)+1}$

$20\log|G| = 20\log\frac{1}{8} + 20\log\omega$

$-20\log\left[\left(1-\frac{\omega^2}{400}\right)^2 + .25\left(\frac{\omega}{20}\right)^2\right]^{1/2}$

$20\log\frac{1}{8} = -18$; $20\log\omega = +18$ at $\omega = 8$

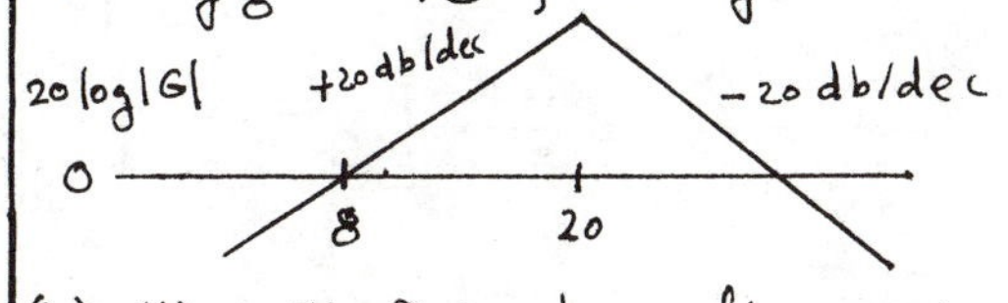

(a)

(b) $r(t) = u(t)$; $R(s) = \frac{1}{s}$ $\lim_{s\to 0} G(s) = 0$

$e_{ss}(t) = \lim_{s\to 0}\frac{s\frac{1}{s}}{1+0} = 1$

(b) $R(s) = \frac{1}{s^2}$: $e_{ss}(t) = \lim_{s\to 0}\frac{s\frac{1}{s^2}}{1+0} = \infty$

4. $GH\Big|_{s\to 0} = \frac{5}{s^2}$ $\qquad \frac{1}{1+GH}\Big|_{s\to 0} = \frac{1}{1+\frac{5}{s^2}} = \frac{s^2}{5}$

UNIT STEP : $R = \frac{1}{s}$

$$e_{ss} = \lim_{s\to 0} s\frac{1}{s}\frac{s^2}{5} = 0$$

UNIT RAMP : $R = \frac{1}{s^2}$

$$e_{ss} = \lim_{s\to 0} s\frac{1}{s^2}\frac{s^2}{5} = 0$$

UNIT parabola : $R = 2/s^2$

$$e_{ss} = \lim_{s\to 0} s\frac{2}{s^2}\frac{s^2}{5} = \frac{2}{5}$$

5.(a) $[s^4 + 30s^3 + 200s^2] + [40s + 40] = 0$

s^4	1	200	40
s^3	30	40	
s^2	$198\frac{2}{3}$	40	
s^1	33.96	0	
s^0	40		

$$b_3 = \frac{\begin{vmatrix} 1 & 200 \\ 30 & 40 \end{vmatrix}}{-30} = 198\tfrac{2}{3}$$

$$b_5 = \frac{\begin{vmatrix} 1 & 40 \\ 30 & 0 \end{vmatrix}}{-30} = 40$$

$$C_3 = \frac{\begin{vmatrix} 30 & 40 \\ 198\frac{2}{3} & 40 \end{vmatrix}}{-198\frac{2}{3}} = 33.96$$

$$d_3 = \frac{\begin{vmatrix} 198\frac{2}{3} & 40 \\ 33.96 & 0 \end{vmatrix}}{-33.96} = 40$$

system is stable

(b) $[s^3 + s^2 + s] + [s + 8] = 0$

s^3	1	2
s^2	1	8
s^1	-6	0
s^0	8	

2 sign changes

$$C_3 = \frac{\begin{vmatrix} 1 & 2 \\ 1 & 8 \end{vmatrix}}{-1} = -6$$

$$d_3 = \frac{\begin{vmatrix} 1 & 8 \\ -6 & 0 \end{vmatrix}}{6} = 8$$

system unstable with 2 positive roots

6. $G = \frac{K}{\tau s+1} = \frac{K}{\frac{s}{4}+1}$

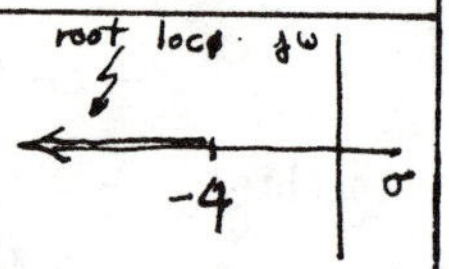

$$1+G = \frac{\frac{s}{4}+1+K}{\frac{s}{4}+1} = \frac{1+K}{\frac{s}{4}+1}\left[\frac{s}{4(1+K)}+1\right]$$

$$\frac{1}{4(1+K)} = .1 \qquad (1+K) = 2.5$$

$$K = 1.5$$

7.

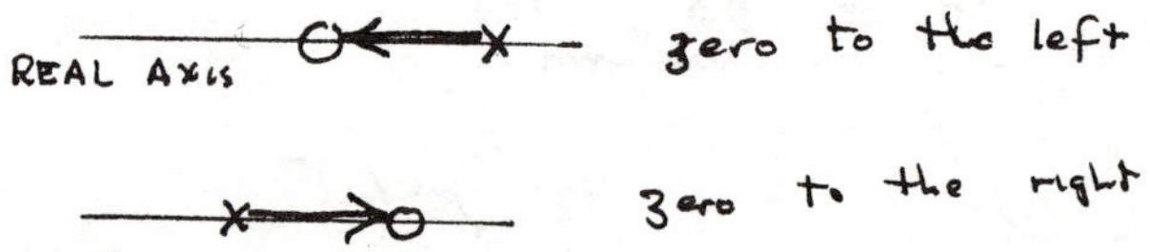

8. eq 11.80 $r = \sqrt{(-9)(-9+3)} = 3\sqrt{6}$

circle :

$$(\sigma+9)^2 + \omega^2 = r^2 = 54$$

for $\zeta = .707$: $\omega = -\sigma$

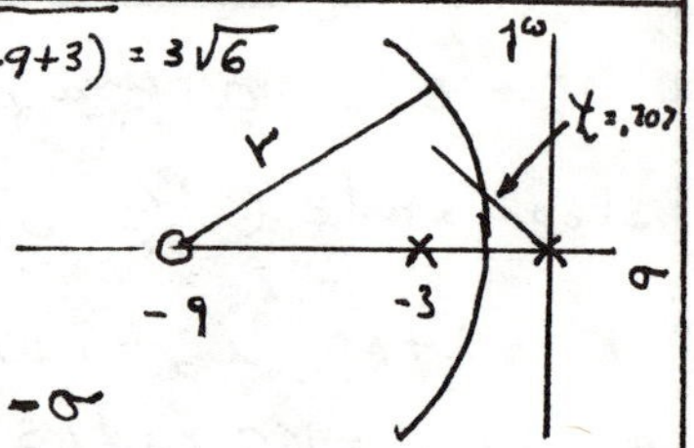

$$\sigma^2 + 18\sigma + 81 + \sigma^2 = 54 \; ; \; \sigma^2 + 9\sigma + 13.5 = 0$$

$$\sigma = -1.902 \; , \; -7.098$$

(a) $s = -1.902(1-j)$ $\qquad 1 + \frac{K(s+9)}{s(s+3)} = 0$

$K = .8038$ $\qquad s^2 + (3+K)s + 9K = 0$

$$K = -\frac{s^2+3s}{s+9}$$

$$\frac{G}{1+GH} = \frac{.8038(s+9)}{s^2 + 3.8038s + 7.235}$$

$$C(s) = \frac{RG}{1+GH} = \frac{.8038(s+9)}{s[(s+1.9019)^2 + 1.9019^2]}$$

$$= \frac{1}{s} - \frac{s+3}{(s+1.9019)^2 + 1.9019}$$

$$= \frac{1}{s} - \frac{(s+1.9019) + 1.098}{(s+1.9019)^2 + 1.9019}$$

table 1.1

$$c(t) = 1 - e^{-1.9t}(\cos 1.9t + .91 \sin 1.9t)$$

(b) $s = -7.098(1-j)$ $\qquad K = 11.20$

$$C(s) = \frac{11.20(s+9)}{s[(s+7.098)^2 + 7.098^2]}$$

$$= \frac{1}{s} - \frac{s+3}{(s+7.098)^2 + 7.098^2}$$

$$= \frac{1}{s} - \frac{(s+7.098) - 4.098}{(s+7.098)^2 + 7.098^2}$$

$$c(t) = 1 - e^{-7.1t}[\cos 7.1t - .58 \sin 7.1t]$$

9. $\dfrac{K(s+2)}{s^2(s+10)(s+20)}$

Centroid: $\dfrac{-10-20+2}{4-1} = -9.33$

$\angle$asymptotes: $\dfrac{360n}{4-1} + 180°$

$= 300°,\ 420° \to 60°$

and, $540° \to 180°$

Assymptote equation

$\omega = \tan 60° (\sigma + 9.33)$

at $\sigma = 0$, $\omega = 9.33 \tan 60° = 16.17$

ROUTH-HURWITZ

$[s^4 + 30s^3 + 200] + [Ks + 2K] = 0$

s^4	1	200	2K
s^3	30	k	
s^2	$\frac{6000-K}{30}$	2K	
s^1	$\frac{4200-K}{6000-K}k$	0	
s^0	2K		

$c_3 = \dfrac{\begin{vmatrix} 1 & 200 \\ 30 & K \end{vmatrix}}{-30} = \dfrac{6000-K}{30}$

$c_1 = \dfrac{\begin{vmatrix} 200 & 2K \\ K & 0 \end{vmatrix}}{-K} = 2K$

$d_3 = \dfrac{\begin{vmatrix} 30 & K \\ \frac{6000-K}{30} & 2K \end{vmatrix}}{-\frac{6000-K}{30}} = \dfrac{4200-K}{6000-K}K$

$K = 4200$

Assymptote value $\omega = 16.17$

$K = \dfrac{-s^2(s+10)(s+20)}{s+2}\Big|_{s=j16.17} = 7.836\angle 64°$

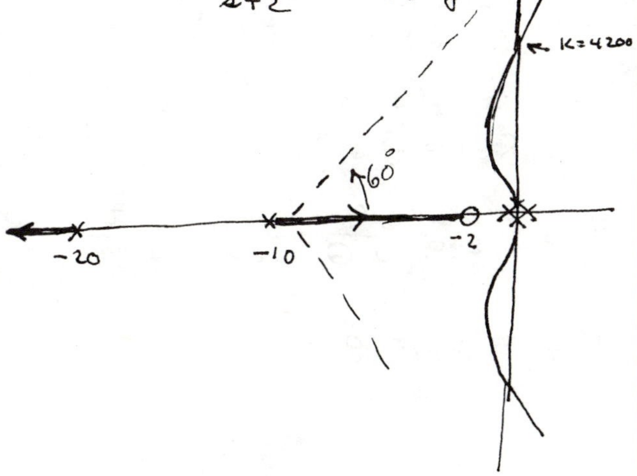

10. $\dfrac{C}{R} = \dfrac{5s^2 + 16s + 25}{15s^4 + 225s^3 + 10s^2 + 16s + 25}$

$= \dfrac{(.333s^2 + 1.067s + 1.667)X_1}{(s^4 + 15s^3 + .667s^2 + 1.067s + 1.667)X_1}$

$R = X_1^{(4)} + 15X_1^{(3)} + .667X_1'' + 1.067X_1' + 1.667X_1$

$C = .333X_1'' + 1.067X_1' + 1.667X_1$

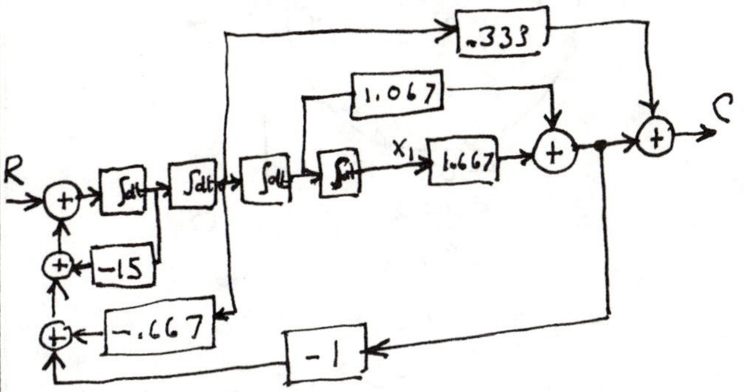

CONCENTRATES

1. $G = \dfrac{20}{s(s+10)}$; $H = (1 + k_T s)$

$\dfrac{G}{1+GH} = \dfrac{20}{s^2 + (10 + 20k_T)s + 20} = \dfrac{20}{s^2 + 2\zeta\omega_n s + \omega_n^2}$

$\omega_n = \sqrt{20}$ $\qquad 2\zeta\omega_n = (10 + 20k_T)$

$\zeta = \dfrac{1}{\sqrt{2}}$ $\qquad 2\sqrt{10} = 10 + 20k_T$

$k_T = -.184$

2.(a) $(s+1)(s+2)(s+10) + 80 = 0$ for stability

$s^3 + 13s^2 + 32s + 100 = 0$

ROUTH HURWITZ

s^3	1	32
s^2	13	100
s^1	24.3	0
s^0	100	

$c_3 = \dfrac{\begin{vmatrix} 1 & 32 \\ 13 & 100 \end{vmatrix}}{-13} = 24.3$

system is stable

$\dfrac{G}{1+GH} = \dfrac{80}{s^3 + 13s^2 + 32s + 100}$ (a)

(b) $\dfrac{G}{1+GH} = \dfrac{80(s+10)}{s^3 + 13s^2 + 32s + 100}$

stability is the same.

the output in (b) will be 10 times that of (a) $C/R|_{s=0} = \frac{80}{100} = .8$

(b) $C/R|_{s=0} = \dfrac{80(10)}{100} = 8$

3. the root-locus is a circle with center at $s = -5$ and with radius $\sqrt{-5(-5+1)} = \sqrt{20}$ [eq. 11.80]

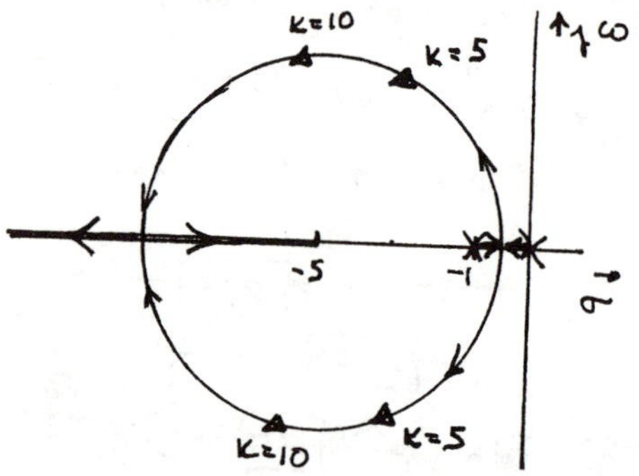

at $K = 5$

$$\frac{5s+25}{s^2+s} + 1 = 0$$

$$s^2 + 6s + 25 = 0$$

$$s = -3 \pm j4$$

at $K = 10$

$$\frac{10s+50}{s^2+s} + 1 = 0$$

$$s^2 + 11s + 50 = 0$$

$$s = -5.5 \pm j4.44$$

4. $c(t) = 20(1 - e^{-5t})$ $\quad e(t) = 10^{-2}u(t)$

$C(s) = \frac{20}{s} - \frac{20}{s+5}$ (table 1.1) $E(s) = \frac{10^{-2}}{s}$

$$G(s) = \frac{C(s)}{E(s)} = \frac{100}{s(s+5)} \times \frac{s}{10^{-2}} = \frac{10^4}{s+5}$$

$$\left.\frac{G}{1+GH}\right|_{s=0} = 10 = \frac{.2\times10^4}{1+.2\times10^4 H} \qquad H = .1$$

$$\frac{G}{1+GH} = \frac{\frac{10^4}{s+5}}{1+\frac{10^3}{s+5}} = \frac{10^4}{s+10^3} = \frac{C}{R}$$

for $R = 1/s$

$$C = \frac{10^4}{s(s+10^3)} = \frac{10}{s} - \frac{10}{s+10^3}$$

$$c(t) = 10\left[1 - e^{-10^3 t}\right]$$

5.

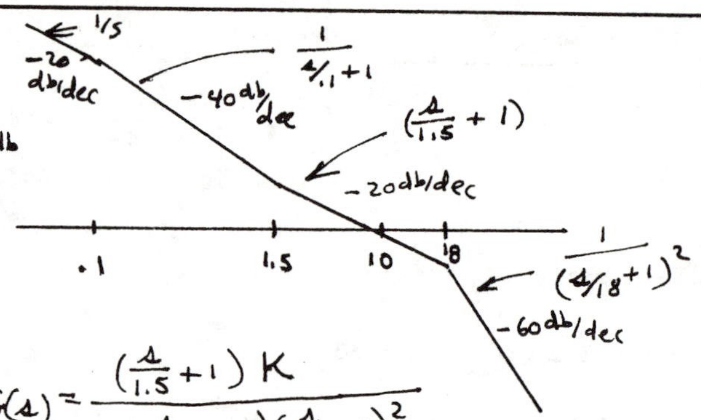

$$G(s) = \frac{\left(\frac{s}{1.5}+1\right)K}{s\left(\frac{s}{.1}+1\right)\left(\frac{s}{18}+1\right)^2}$$

$$|G(j10)| = 1 \;:\; K = \frac{10\left|1+j\frac{10}{.1}\right| \times \left|1+j\frac{10}{18}\right|^2}{\left|1+j\frac{10}{1.5}\right|}$$

$$K = 194$$

$$G = G_p G_c = \frac{194\left(\frac{s}{1.5}+1\right)}{s\left(\frac{s}{.1}+1\right)\left(\frac{s}{18}+1\right)^2} \quad (a)$$

$$G_c = \frac{G}{G_p} = \frac{194\left(\frac{s}{1.5}+1\right)}{s\left(\frac{s}{.1}+1\right)\left(\frac{s}{18}+1\right)^2} \cdot \frac{s(s+2)(s+10)}{20}$$

$$= \frac{194\left(\frac{s}{1.5}+1\right)\left(\frac{s}{2}+1\right)\left(\frac{s}{10}+1\right)}{\left(\frac{s}{18}+1\right)^2\left(\frac{s}{.1}+1\right)}$$

for $r(t) = (5+3t)u\,t$

$$R(s) = \frac{5}{s} + \frac{3}{s^2}$$

$$e_{s.s.} = \lim_{s\to0} \frac{s\,R(s)}{1+G} = \lim_{s\to0} \frac{5+\frac{3}{s}}{1+\frac{194}{s}}$$

$$e_{s.s.} = \frac{3}{194} = .0155$$

TIMED

1. $R = (s^4 + 22s^3 + 41s^2 + 20s)\,x$

$C = (s^2 + 100)\,x$

$$\frac{C}{R} = \frac{s^2+100}{s(s^3+22s^2+41s+20)}$$

$$\begin{array}{r} s+20 \\ s^2+2s+1 \,\big)\, \overline{s^3+22s^2+41s+20} \\ s^3+2s^2+s \\ \hline 20s^2+40s+20 \end{array}$$

s^2+2s+1 ↑ NATURAL DAMPING at 1 rad/sec

$$\frac{C}{R} = \frac{(s+10j)(s-10j)}{s(s+20)(s+1)^2} \quad (a)$$

(b)
$$G = \frac{\frac{C}{R}}{1-\frac{C}{R}} = \frac{s^2+100}{s^2+22s^3+40s^2+20s-400}$$

↑ which is unstable

(c) Maintaining the pole at $s=-20$: and $\omega_n=1$:

$$\begin{array}{r} s^2+2\zeta s+1 \\ \times(s+20) \\ \hline s^3+(20+2\zeta)s^2+(1+40\zeta)s+20 \end{array}$$

↑ −22 box ↑ −41 box

both the −22 box and the −41 box affect ζ, however a small change in the −22 box indicates a large change in ζ. Thus a small change in the −22 box results in a large change in ζ.

2. $[s^3 + 16s^2 + 72s + 126] + [K] = 0$

ROUTH-HORWITZ.

s^3	1	72
s^2	16	126+K
s^1	C_3	
s^0	d_3	

$$C_3 = \frac{126+K-16\times 72}{-16} = \frac{1026-K}{16}$$

$$d_3 = 126+K$$

for positive K : $K < 1026$

negative K : $K > -126$

$126 < K < 1026$

3. first reduce the inner loop:

$$\frac{\frac{100}{s(s+10)}}{1+\frac{100}{s(s+10)}} = \frac{100}{s^2+10s+100}$$

$$G = \frac{1000K/100}{\left(\frac{s}{10}\right)^2 + \frac{s}{10} + 1}$$

$\omega_n = 10$

$2\zeta\omega_n = 10$

$\zeta = .5$

roots: $s = -5 \pm \sqrt{75}$

$= -5 \pm j8.66 = 10\angle \pm 120°$

root loci

$s^2 + 10s + 100 + 1000K = 0$

$s = -5 \pm j\sqrt{100+1000K-25}$

K	.05	.1	.15	.2	.25
$\sqrt{75+1000K}$	11.2	13.2	15	16.6	18

.5	.75	1
24	28.7	32.8

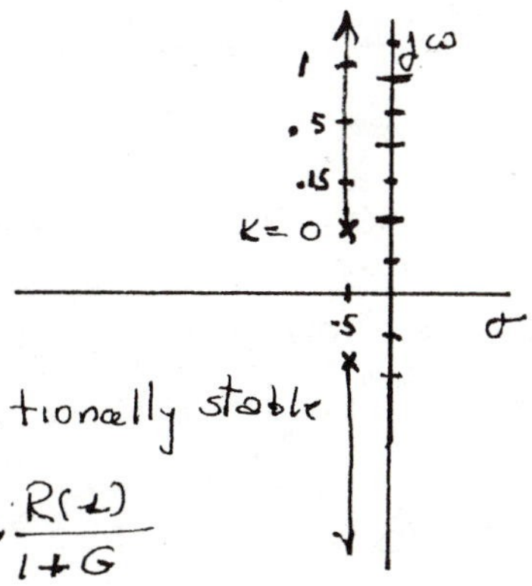

(a)

(b) it is unconditionally stable

(C) $e_{ss} = \lim_{s\to 0} s\,\frac{R(s)}{1+G}$

$R(s) = \frac{1}{s}$ normalized

$\lim_{s\to 0} G(s) = 10K$

$$e_{ss} = \lim_{s\to 0} \frac{s\,\frac{1}{s}}{1+10K} = \frac{1}{1+10K}$$

4. Root Locus: $s^3+2s^2+4s+K=0$

at $K=0$ (open loop poles)

$s(s^2+2s+4)=0 \quad \therefore s_0=0$

$s_1, s_2 = -1 \pm j\sqrt{3}$

(a) there are no breakaway points, as only one pole is on the real axis.

(b) for roots on the imaginary axis

$s^3+2s^2+4s+K=(s^2+\omega^2)(s+P)$

using long division

$$\begin{array}{r} s+2 \\ s^2+\omega^2 \overline{)\, s^3+2s^2+4s+K} \\ s^3 \qquad \omega^2 s \\ \hline 2s^2+(4-\omega^2)s+K \\ 2s^2 \qquad +\omega^2 \\ \hline (4-\omega^2)s+(K-\omega^2) \end{array}$$

for this to have no remainder

$4-\omega^2=0$ and $K-\omega^2=0$

or $\omega=2$, $K=4$ (b)

(c) At $K=0$ $s^2+2s+4 = s^2+2\zeta\omega_n s+\omega_n^2$

$\omega_n=2 \quad \zeta=.5$

as K increases, ζ decreases, so a point where $\zeta=.707$ does not occur for $K>0$.

ROOT LOCUS

CENTROID: $\frac{-1-1}{3}=-2/3$

Asymptotes: $180°, 300°, 420°=60°$

Lemergence:

$180+120+90=390$

$\theta=390-360=30°$

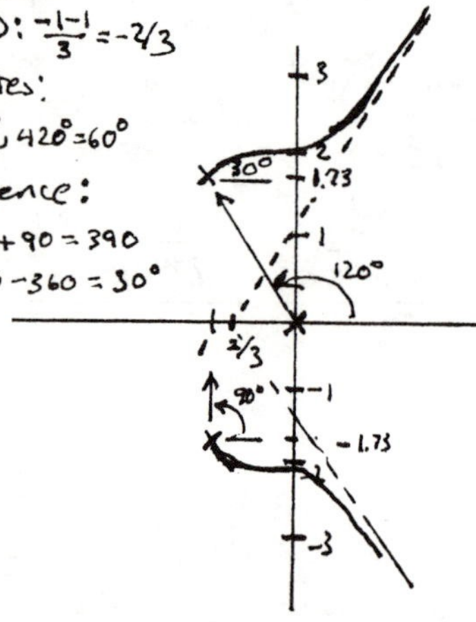

5.

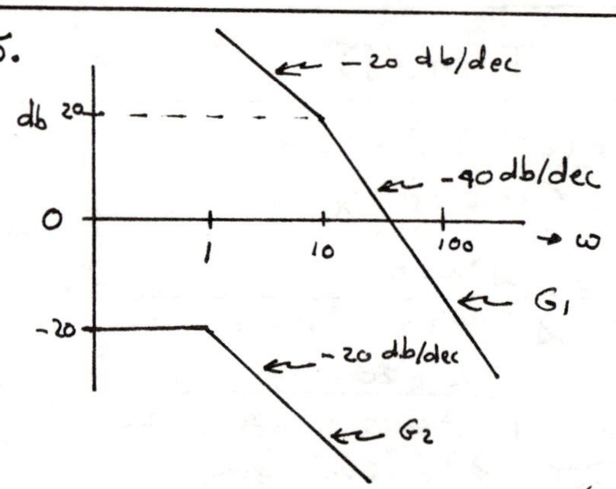

High-frequency information indicates that G_1 has 2 more poles than zeros, and that G_2 has one more pole than zeros. This indicates "minimum phase systems," i.e. no zero's in the right-half plane. This allows the assymptotes to define the transfer functions:

$G_1 = \frac{K_1}{s(s+10)} \qquad |G_1(j10)| = 10$

then $K_1 = 10 \times |j10| \times |j10+10| = 10^3\sqrt{2}$

$G_2 = \frac{K_2}{(s+1)} \qquad$ below $j1$, $|G_2| = .1$

$\therefore K_2 = .1$

$$G = G_1 G_2 = \frac{100\sqrt{2}}{s(s+1)(s+10)}$$

ROUTH-HURWITZ

$[s^3+11s^2+10s]+[100\sqrt{2}]=0$

s^3	1	10
s^2	11	$100\sqrt{2}$
s^1	-2.85 sign change	
s^0	141.4 sign change	

$c_3 = \frac{\begin{vmatrix} 1 & 10 \\ 11 & 100\sqrt{2} \end{vmatrix}}{-11} = -2.85$

$d_3 = 100\sqrt{2}$

the system is unstable, with two poles in the right-hand plane.

WARMUPS

1. CCR = 0 (because $h_{cc} = 0$)

FCR = 0.52 as in ex. 12.1

RCR = 5 x (24 − 2.5)/24 = 4.5

2. $CCR = \frac{5 \times 12}{24} = 2.5$

FCR = 0.52 as in ex 12.1

$RCR = \frac{5 \times (24 - 2.5 - 12)}{24} = 1.98$

3. Ceiling reflectance = 90% → ρ_c

Wall reflectance = 75% → ρ_w

From table 12.5 under base reflectance 90% (ceiling = base), 2d & 3d columns at CR = 2.5

ρ_w =	80%	70%
ρ_{cc} =	75%	68%

interpolating $\frac{\rho_{cc} - 68\%}{75\% - 68\%} = \frac{75\% - 70\%}{80\% - 70\%}$

effective ceiling Refl: ρ_{cc} = 71.5%

4. From Table 12.1 for RCR = 2, ceiling cavity reflectance of 80% extrapolate from wall reflectances of 50% and 30% to obtain reflectance of 75% of CU = .845

Repeat for ceiling cavity reflectance of 50% to obtain CU = .763

Interpolate between these values to obtain value at ceiling reflectance of 71.5% CU = .822

5. Total lumens = $\frac{100 \text{ f.c} \times 40' \times 60'}{CU} = \frac{240,000}{.822}$

292,000 lumens

6. LLF = .98 x .84 x .94 x .95 = .735

7. $N_{luminaires} = \frac{100 \times 40 \times 60}{2 \times 3000 \times .822 \times .735}$

= 66.2 → 67 minimum

8. With rows in the 40′ direction and 52″ in length

$N/row = N_r = \frac{40 \times 12}{52} \rightarrow 9$

$N \text{ of Rows} = Rows = \frac{67}{9} \rightarrow 8$

distance between Rows = L_R

$L_R = \frac{12 \times 60}{8 - 1 + 2/3} = 93.9 \rightarrow 94''$

distance from last luminaire in row to wall (40′ direction) = L_e

$L_e = \frac{1}{2}(40 \times 12 - 9 \times 52) = 6.0''$

distance to wall from last row = L_w

$L_w = \frac{1}{2}(12 \times 60 - 7 \times 94) = 31.0''$

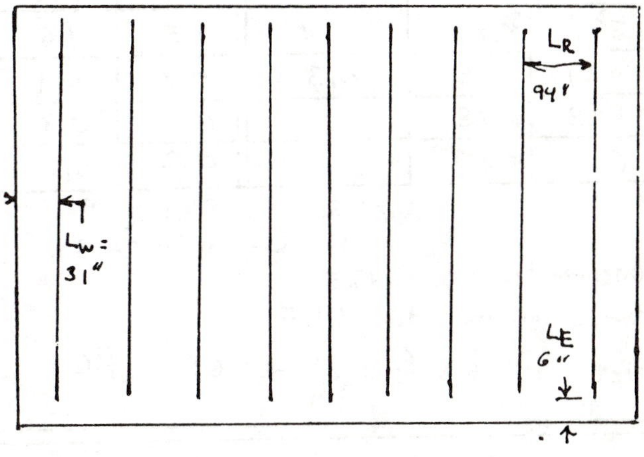

9. RCR ≈ 2, ρ_{cc} = 71.5%, ρ_w = 75%

At B1: RPM = .51; At B2: RPM = .6

interpolating: at B1.5: RPM = .555

From the bottom of Table 12.7

RCC = 2	WALL (LC_w)				CEILING (LC_{cc})			
Ceiling Refl ρ_{cc}	80%		50%		80%		50%	
Wall Refl. ρ_w	50%	30%	50%	30%	50%	30%	50%	30%
LC	.232	.127	.209	.115	.222	.190	.130	.113
ρ_w = 75%	.363		.327		.262		.151	
ρ_{cc} = 71.5%	LC_w = .353				LC_{cc} = .231			

EQ 2.11 RRC = .285

72 LUMINAIRES: $FC_{rh} = \frac{2 \times 72 \times 3000 \times .704 \times .285}{24.1} = 36.1$ FC

10.

END VIEW (60')

SIDE VIEW (40')

$\alpha_2 = 74.2°$

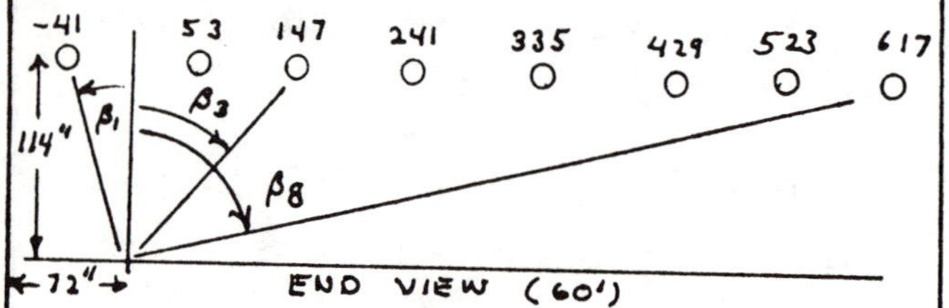

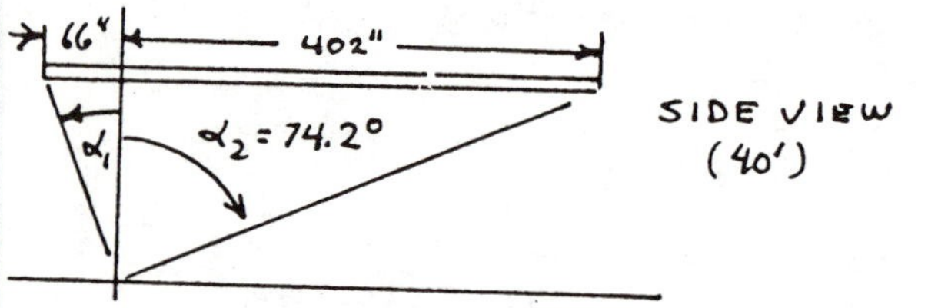

from table 12.7 : * = interpolated values

LATERAL DISTANCE (INCHES)	β (degrees)	Footcandles for $\alpha_1 = 30°$	FOOTCANDLES for $\alpha_2 = 70°$	for $\alpha_2 = 80°$
-41	*20	24.1	39.7	40.3
53	25	21.6	36.4	36.9
147	*52	5.7	11.9	12.3
241	65	1.1	3.2	3.5
335	*71	0.4	1.5	1.6
429	75	0.3	0.3	0.4
523	*78	0	0.2	0.3
617	*80	0	0.2	0.3
		53.2	93.4	95.6

interpolate 94.3

uncorrected f.c. = 147.5

Correction including LLF

Net = $147.5 \times \frac{6}{9.5} \times .704 = 65.6$ f.c.

CONCENTRATES

6. Required f.c. = $90 \times 56 \times 24 = 121,000$

Assuming Ceiling mounted 2 T 12 luminaires:

$RCR = 5 \times (14 - 2.5) \times \frac{24 + 56}{24 \times 56} = 3.42$

CU from table 12.1 with 50% wall reflect.

RCR	for ceiling reflectance 80%	50%	EXTRAPOLATE for 90%
3	.68	.61	.70
4	.60	.54	.2

interpolate for RCR = 3.42 ⟶

$$CU = .67$$

Assume Enclosed Luminaire
Page 12.4 table 12.2 LDD = 0.88

from page 12-5 (table 12.3)
use high output F48T12 lamps
with multiplying factor of 1.03 (see footnote):

Initial lumens/lamp = IL = $4200 \times 1.03 = 4326$

also LLF = .80

Number of luminaires (2 lamps/luminaire)

$$N_L = \frac{121000}{4326 \times 2 \times .67 \times .88 \times .80} = 29.7 \rightarrow 30$$

24' dimension - max luminaires/row

Luminaires/row = $\frac{24' \times 12}{52"} = 5.54 \rightarrow 5$

#rows along 56' dimension = $\frac{30}{5} = 6$

56' dimension - max luminaires/row

Luminaires/row = $\frac{56' \times 12}{52"} = 12.92 \rightarrow 12$

rows along 24' dimension = $\frac{30}{12} = 2.5 \rightarrow 3$

Place rows in short dimension as fewer luminaires are needed

$N_{rows} = 6$, end spacing $\frac{24 \times 12 - 5 \times 52}{2} = 14"$

Row spacing = $\frac{56 \times 12}{6 - 1 + \frac{2}{3}} = 118.5"$

spacing to wall = $\frac{56 \times 12 - 5 \times 118.5}{2} = 39\frac{3}{4}"$

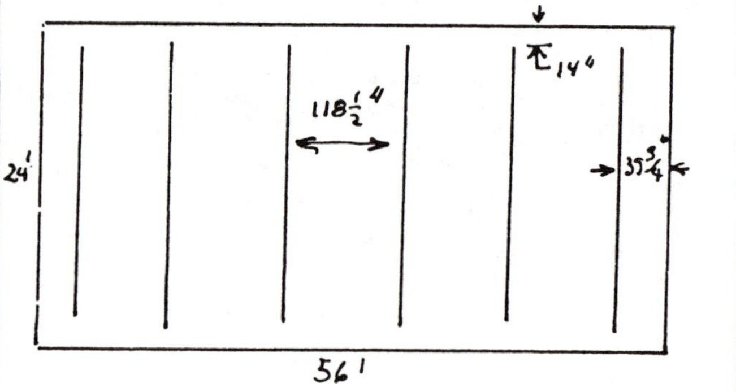

2. $N_{luminaires} = \dfrac{100 \times 96 \times 180}{4 \times 3100 \times .75 \times .62} = 299.7$

then 300 8' luminaires are needed

$8'\text{luminaires/row} = \dfrac{180 \times 12}{100} = 21.6$

then each row consists of

21 - 8' and one 4' luminaire = 21.5

$\#\ \text{rows} = \dfrac{300}{21.5} = 13.95 \rightarrow 14 \text{ rows}$

$\text{END SPACING} = \dfrac{180 \times 12 - 21.5 \times 100}{2} = 5 \text{ inches}$

$\text{Row spacing} = \dfrac{96 \times 12}{14 - 1 + 2/3} = 84\tfrac{1}{4}''$

spacing from last row to WALL

$\dfrac{96 \times 12 - 84\frac{1}{4} \times 13}{2} = 28\tfrac{3}{8}''$

(a) 294 - 8' and 14 - 4' luminaires

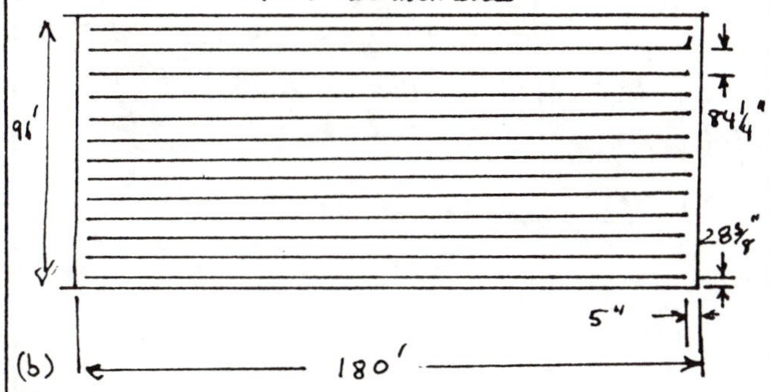

$P = (285 \times 21 + 190) \times 14 = 86.45 \text{ KW}$

$W = 86.45 \times 17 \frac{\text{hrs}}{\text{day}} \times 5 \frac{\text{days}}{\text{week}} \times 50 \frac{\text{weeks}}{\text{yr}} = 367412 \text{ KW hrs}$

(c)

$\text{cost/yr} = 367412 \times .015 = \$5{,}511.19$

3. table 12.8 : MMI = 2 f.c.

P 12-18 : LDD = .92

example 12.1 : LLD = .91

for MH = 30' and 6' overhang:

$CU(SS) = .35$, $CU(HS) = .02$ } $CU = .37$

spacing for staggered or opposite:

$\dfrac{2 \times 1 \times 25500 \times (.91 \times .92) \times .37}{2 \times 40} = 197.5'$ (eq 12.15)

TIMED

1.

	A	B
Luminaires	$\dfrac{30 \times 20000}{1 \times 2600 \times .78 \times .72} = 411$	$\dfrac{30 \times 20000}{1 \times 7700 \times .78 \times .72} = 139$
Initial Cost	$411 \times 75 = \$30{,}825$	$139 \times 162.50 = 22{,}875$
Energy cost/yr	$411 \times .15\text{KW} \times 3200 \times .075 = \$14{,}797.50$	$139 \times .175 \times 3200 \times .075 = \$5{,}838.00$
Replacement cost/yr	$411 \times \dfrac{3200}{750} \times 4.35 = \$7{,}628.16$	$139 \times \dfrac{3200}{24000} \times 10 = \$185.33$
Annual Cost	\$ 22,425.66	\$ 6,020.33

2. A scheme with 4 lamps/pole in the middle of the lot, 1 lamp per pole on the corners, and 2 lamps per pole along the edges will give reasonably uniform lighting.

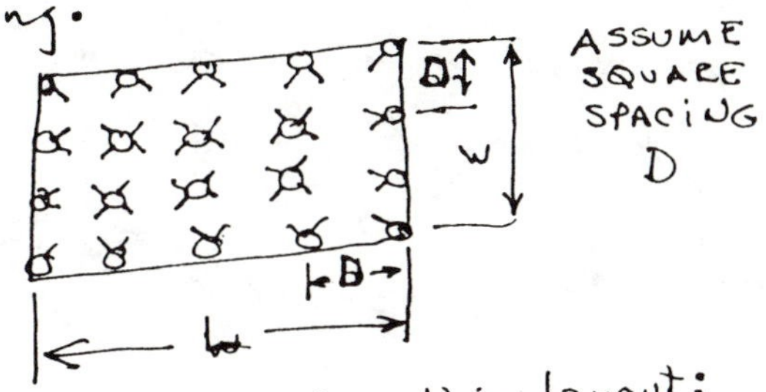

Number of lamps for this layout:

Corner : 4

edges : $2 \times (2(\frac{L}{D} - 1)) + 2 \times (2(\frac{W}{D} - 1))$

MIDDLE : $4 \times (\frac{L}{D} - 1)(\frac{W}{d} - 1)$

$N_L = 4\dfrac{L \times W}{D^2}$ or $D = 2\sqrt{\dfrac{L \times W}{N_L}}$

$N_L = \dfrac{6.25 \times 360 \times 420}{6500 \times .65 \times .74 \times .48} = 63 \text{ lamps}$

then $D = 2\sqrt{\dfrac{360 \times 420}{63}} = 98'$ maximum

$\dfrac{W}{D} = \dfrac{360}{98} = 3.67 \rightarrow 4 : 5 \text{ poles across}$

$\dfrac{L}{D} = \dfrac{420}{98} = 4.29 \rightarrow 5 : 6 \text{ poles lengthwise}$

2 CONTINUED

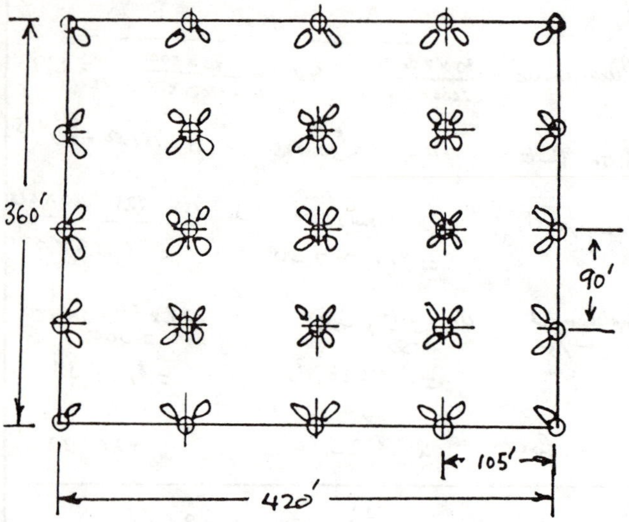

$N_{Lamps} = 4 + 12 \times 2 + 9 \times 4 = 64$

EACH LAMP REQUIRES $I = \frac{(1000+25)\,W}{220\,V \times .8}$

assuming 220 V line-neutral and a powerfactor of .8. Then each lamp must have a de-rating of 0.8 to meet NEC, so

$$I_{rated}/lamp = \frac{1025}{220 \times .8 \times .8} = 7.28$$

BEAM ANGLE

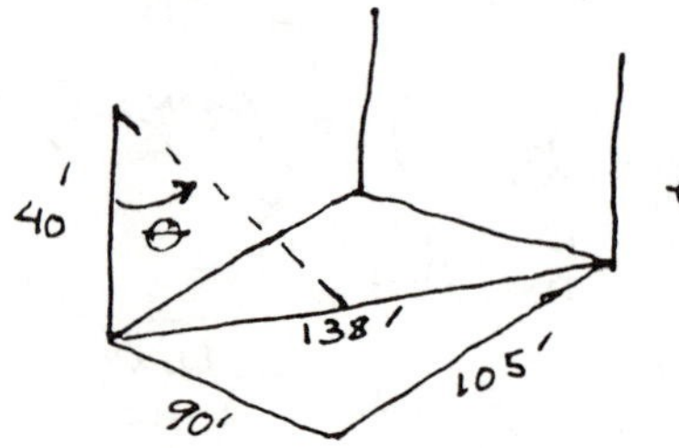

$\theta = \tan^{-1} \frac{138/2}{40}$

$= 60°$

22 lamps are required per phase so total rating must be 160 A the next higher rated breaker is 175 A, so the breaker must be rated: 3-phase 175 A, $220\sqrt{3} = 381$ V with a surge current of 280 A

This will feed 3 parallel load circuits

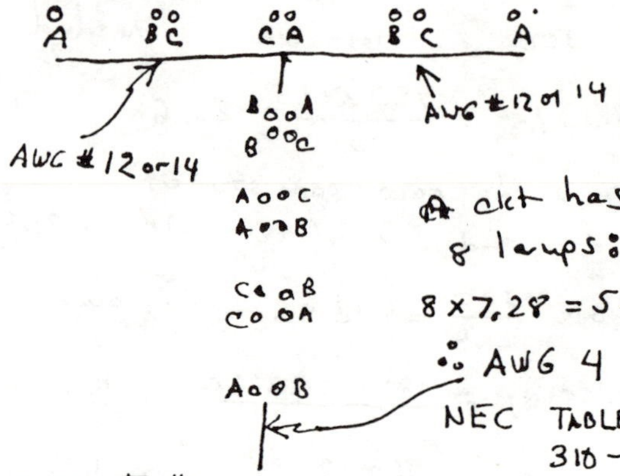

A ckt has 8 lamps:

$8 \times 7.28 = 58.24$

∴ AWG 4

NEC TABLE 310-6

CIRCUIT #1

CIRCUIT #2 (#3 is the same, mirror)

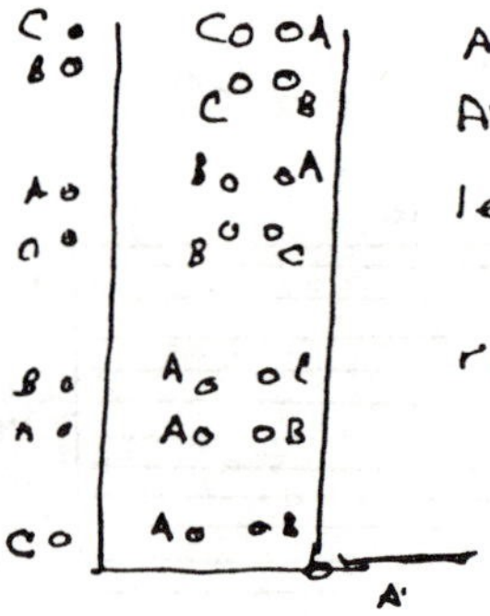

All phases have 7 loads

AWG #6 feed

left branch: 114.6 A

2 lamps: AWG #12 or 14

right branch: 36.4 A

5 lamps: AWG #8

HOMEWORK SOLUTIONS: ENGINEERING ECONOMICS

WARM-UPS

1 $F = 1000\,(F/P, 6\%, 10)$

$= 1000\,(1.7908) = 1790.80$

2 $P = 2000\,(P/F, 6\%, 4)$

$= 2000\,(.7921) = 1584.20$

3 $P = 2000\,(P/F, 6\%, 20)$

$= 2000\,(.3118) = 623.60$

4 $500 = A\,(P/A, 6\%, 7)$

$= A\,(5.5824)$

$A = \frac{500}{5.5824} = 89.57$

5 $F = 50\,(F/A, 6\%, 10)$

$= 50\,(13.1808) = 659.04$

6 EACH YEAR IS INDEPENDENT

$\frac{200}{1.06} = 188.68$

7 $2000 = A + A(F/A, 6\%, 4)$

$= A + A\,(4.3746)$

$A = 372.12$

8 $F = 100\left[(F/P, 6\%, 10) + (F/P, 6\%, 8) + (F/P, 6\%, 6)\right]$

$= 100\,(1.7908 + 1.5938 + 1.4185)$

$= 480.31$

9 $r = .06$

$\phi = \frac{.06}{12} = .005$

$N = 5(12) = 60$

$F = 500\,(1.005)^{60} = 674.43$

10 $120 = 80\,(F/P, ?, 7)$

$(F/P, ?, 7) = \frac{120}{80} = 1.5$

SEARCHING THE TABLES, $? = 6\%$

CONCENTRATES

1 $EUAC = (17000 + 5000)(A/P, 6\%, 5)$

$- (14000 + 2500)(A/F, 6\%, 5) + 200$

$= (22000)(.2374) - (16500)(.1774) + 200$

$= 2495.70$

2 ASSUME THE BRIDGE WILL BE THERE FOREVER.

KEEP OLD BRIDGE

THE GENERALLY ACCEPTED METHOD IS TO CONSIDER THE SALVAGE VALUE AS A BENEFIT LOST (COST).

$EUAC = (9000 + 13000)(A/P, 8\%, 20)$

$- 10{,}000(A/F, 8\%, 20) + 500$

$= (22000)(.1019) - (10{,}000)(.0219) + 500$

$= 2522.80$

REPLACE

$EUAC = 40{,}000(A/P, 8\%, 25)$

$- 15000\,(A/F, 8\%, 25) + 100$

$= (40{,}000)(.0937) - 15000(.0137) + 100$

$= 3642.50$

KEEP OLD BRIDGE

3 $D = \frac{150{,}000}{15} = 10{,}000$

$0 = -150{,}000 + (32000)(1-.48)(P/A, ?, 15)$

$- 7530(1-.48)(P/A, ?, 15)$

$+ 10000(.48)(P/A, ?, 15)$

$150{,}000 = [16640 - 3915.60 + 4800](P/A, ?, 15)$

$(P/A, ?, 15) = \frac{150{,}000}{17524.40} = 8.5595$

SEARCHING THE TABLES, $i = 8\%$

4 a) $\frac{1{,}500{,}000 - 300{,}000}{1{,}000{,}000} = 1.2$

b) $1{,}500{,}000 - 300{,}000 - 1{,}000{,}000$

$= 200{,}000$

5 ANNUAL RENT IS (12)(75) = 900

$F = (14000 + 1000)(F/P, 10\%, 10)$

$+ (150 + 250 - 900)(F/A, 10\%, 10)$

$= 15000(2.5937) - 500(15.9374)$

$= 30936.80$

6 $2000 = 89.30(P/A, ?, 30)$

$(P/A, ?, 30) = \frac{2000}{89.30} = 22.396$

? = 2% PER MONTH

$i = (1.02)^{12} - 1 = .2682$ OR 26.82%

7 SL $D = \frac{500,000 - 100,000}{25} = 16000$

SOYD $T = \frac{1}{2}(25)26 = 325$

$D_1 = \frac{25}{325}(500,000 - 100,000) = 30769$

$D_2 = \frac{24}{325}(400,000) = 29538$

$D_3 = \frac{23}{325}(400,000) = 28308$

DDB $D_1 = \frac{2}{25}(500,000) = 40,000$

$D_2 = \frac{2}{25}(500,000 - 40,000) = 36,800$

$D_3 = \frac{2}{25}(500,000 - 40,000 - 36,800)$

$= 33,856$

8 $P = -12000 + 2000(P/F, 10\%, 10)$

$-1000(P/A, 10\%, 10) - 200(P/G, 10\%, 10)$

$= -12000 + 2000(.3855) - 1000(6.1446)$

$-200(22.8913)$

$= -21951.86$

$EUAC = 21951.86(A/P, 10\%, 10)$

$= 21951.86(.1627) = 3571.56$

9 ASSUME THAT THE PROBABILITY OF FAILURE IN ANY OF THE N YEARS IS 1/N

$EUAC(9) = 1500(A/P, 6\%, 20) + \frac{1}{9}(.35)(1500)$

$+ (.04)(1500)$

$= 1500\left[.0872 + (.35)\left(\frac{1}{9}\right) + .04\right] = 249.13$

$EUAC(14) = 1600\left[.1272 + (.35)\left(\frac{1}{14}\right)\right] = 243.52$

$EUAC(30) = 1750\left[.1272 + (.35)\left(\frac{1}{30}\right)\right] = 243.01$

$EUAC(52) = 1900\left[.1272 + (.35)\left(\frac{1}{52}\right)\right] = 254.47$

$EUAC(86) = 2100\left[.1272 + (.35)\left(\frac{1}{86}\right)\right] = 275.67$

CHOOSE THE 30 YEAR PIPE

10 $EUAC(7) = .15(25000) = 3750$

$EUAC(8) = 15000(A/P, 10\%, 20)$

$+ .10(25000)$

$= 15000(.1175) + .10(25000)$

$= 4262.50$

$EUAC(9) = 20000(.1175) + .07(25000)$

$= 4100.00$

$EUAC(10) = 30,000(.1175) + .03(25000)$

$= 4275$

CHEAPEST TO DO NOTHING

TIMED

1 $EUAC(1) = 10,000(A/P, 20, 1) + 2000$

$- 8000(A/F, 20\%, 1)$

$= 10,000(1.2000) + 2000 - 8000(1.0000) = 6000$

$EUAC(2) = 10,000(A/P, 20, 2) + 2000$

$+ 1000(A/G, 20\%, 2) - 7000(A/F, 20\%, 2)$

$= 10,000(.6545) + 2000 + 1000(.4545)$

$-7000(.4545) = 5818.00$

$EUAC(3) = 10,000(A/P, 20\%, 3) + 2000$

$+ 1000(A/G, 20\%, 3) - 6000(A/F, 20\%, 3)$

$= 10000(.4747) + 2000 + 1000(.8791)$

$- 6000(.2747) = 5977.90$

$EUAC(4) = 10,000(A/P, 20, 4) + 2000$

$+ 1000(A/G, 20\%, 4) - 5000(A/F, 20\%, 4)$

$= 10,000(.3863) + 2000 + 1000(1.2742)$

$- 5000(.1863) = 6205.7$

TIMED #1, CONTINUED

$$EUAC(5) = 10{,}000(A/P,20\%,5) + 2000 + 1000(A/G,20\%,5) - 4000(A/F,20\%,5)$$

$$= 10{,}000(.3344) + 2000 + 1000(1.6405) - 4000(.1344) = 6446.4$$

SELL AT END OF 2ND YEAR

b) FROM EQUATION 13.7,

$$COST = 6000 + .20(5000) + (5000-4000) = 8000$$

2 THE MAN SHOULD CHARGE HIS COMPANY ONLY FOR THE COSTS DUE TO THE BUSINESS TRAVEL:

INSURANCE $300 - 200 = 100$

MAINTENANCE $200 - 150 = 50$

SALVAGE $(1000-500)$ IN 5 YEARS

$$500(A/F,10\%,5) = 500(.1638) = 81.90$$

GASOLINE $\frac{5000(.60)}{15} = 200$

$$EUAC \text{ PER MILE} = \frac{100 + 50 + 81.9 + 200}{5000} = \$.0864$$

a) YES. $\$.10 > \$.0864$ SO IT IS ADEQUATE

b) $$(.10)X = 5000(A/P,10\%,5) + 250 + 200 - 800(A/F,10\%,5) + \frac{X}{15}(.60)$$

$$= 5000(.2638) + 250 + 200 - 800(.1638) + .04X$$

$$.10X = 1637.96 + .04X$$

$$.06X = 1637.96$$

$$X = 27299 \text{ miles}$$

1. Assumptions: $P_L = P_{base} \times pf$

$$P_{base} = \frac{30\,KW}{.8} = 37{,}500$$

$$|I_L| = I_{base}$$

$$|V_L| = V_{base}$$

take V_L = reference = $1\angle 0^\circ$

$$I_L = 1\angle -\cos^{-1}.8$$

$$V_2 = I_L(.005 + j.05) + V_L$$

$$= 1.034 + j.037 = 1.035\angle 2.05^\circ$$

then

$$I_2 = \frac{.5}{V_2}\angle -\tan^{-1}.8 + 2.05 = .483\angle -34.82^\circ$$

$$I_1 = I_L - I_2 = .403 - j.324 = .517\angle -38.78^\circ$$

$$V_1 = (.01 + j.1)I_1 + V_2 = 1.07 + j.074$$

$$S_1 = V_1 I_1^* = .555\angle 42.7^\circ$$

$$= .407 + j.377$$

then $P_1 = .407 \times 37.5\,K = 15.26$ KW

$$pf = \frac{.407}{.555} = .734 \text{ lagging}$$

$$V_1 = |1.07 + j.074| \times 480 = 515\ V_{\ell-\ell}$$

2. Use A as the least-significant variable in Veitch diagram

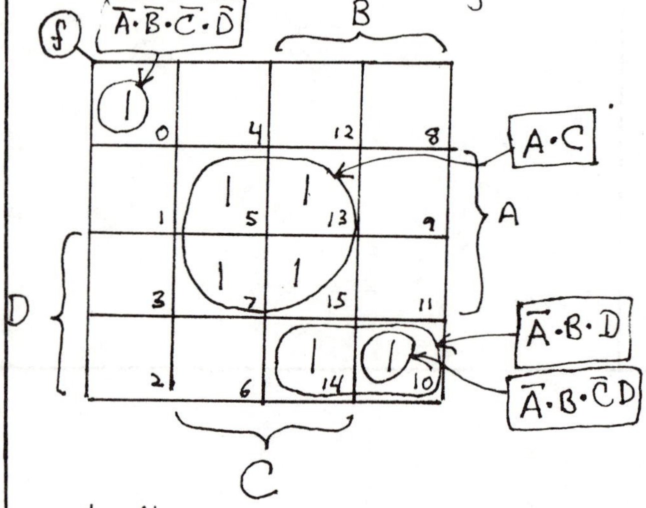

All other entries are zero

Construct truth table with A as the RESIDUE VARIABLE, using the minterm numbers from the Veitch diagram (in lower right-hand corner of each CELL

m	B C D	A	f	R	input #
0	0 0 0	0	1	$\bar{A}$	0
1	0 0 0	1	0		
2	0 0 1	0	0	0	1
3	0 0 1	1	0		
4	0 1 0	0	0	A	2
5	0 1 0	1	1		
6	0 1 1	0	0	A	3
7	0 1 1	1	1		
8	1 0 0	0	0	0	4
9	1 0 0	1	0		
10	1 0 1	0	1	$\bar{A}$	5
11	1 0 1	1	0		
12	1 1 0	0	0	A	6
13	1 1 0	1	1		
14	1 1 1	0	1	1	7
15	1 1 1	1	1		

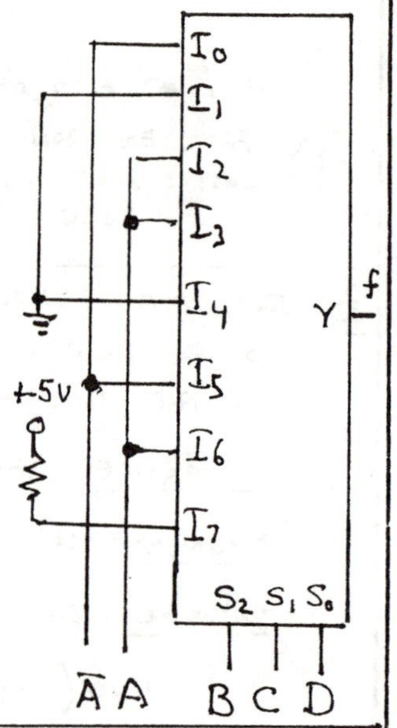

3 see figure 9.35

Region A: Begins with $I_d = I_L = .1\,mA$ at $t=0$.
Ends when $I_d = .4\,mA$ at $t = T_A$.

from graph: $\frac{V_d}{I_d} = \frac{.1}{.0004} = 250\Omega$ (A)

EQUIVALENT CIRCUIT

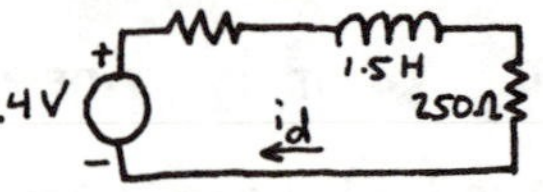

$\tau = 1.5H/850\Omega = 1.765\,mS$

$i_d = i_d(\infty) + (i_d(0) - i_d(\infty))e^{-t/\tau}$

$i_d(0) = .1\,mA$

$i_d(\infty) = \frac{.4}{850}$

then $i_d = .47 \times 10^{-3} + (-.37)e^{\frac{-1700}{3}t} \times 10^{-3}$ (1)

at $t = T_A$, $i_d = .4\,mA$

solve (1) for $t = T_A = 2.93\,mS$

Region C: Begins with $t = T_A$ and $i_d = .4\,mA$.
Ends with $I_d = .1\,mA$ at T_B.

from graph in region C:

$$\frac{V_d - .3}{I_d \quad .1 \times 10^{-3}} = \frac{.6 - .3}{(.4 - .1) \times 10^{-3}} = 10^3$$

$$V_d = .2 + 1000\,I_d$$

EQUIVALENT CIRCUIT

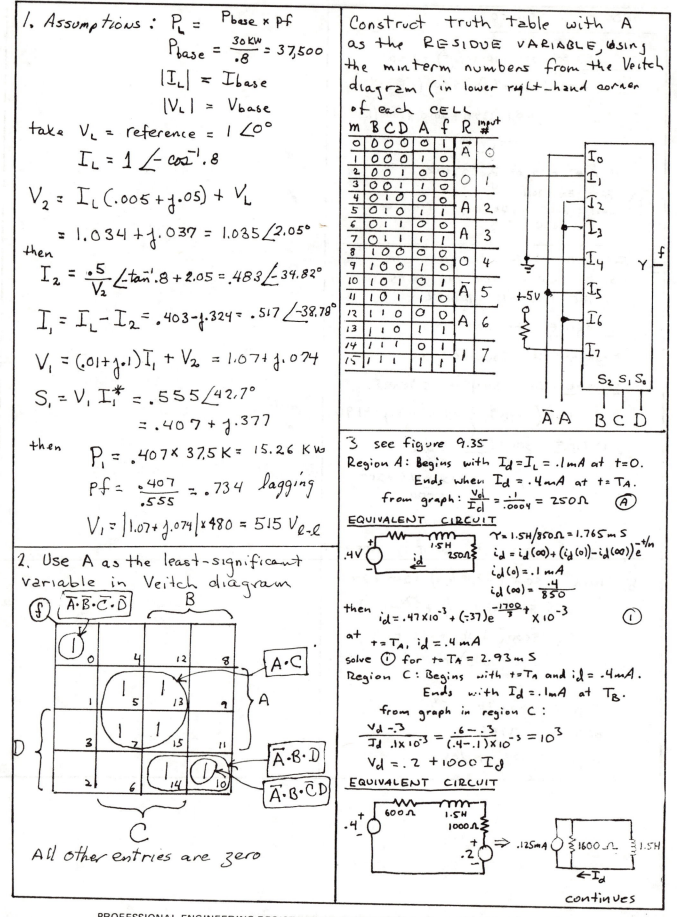

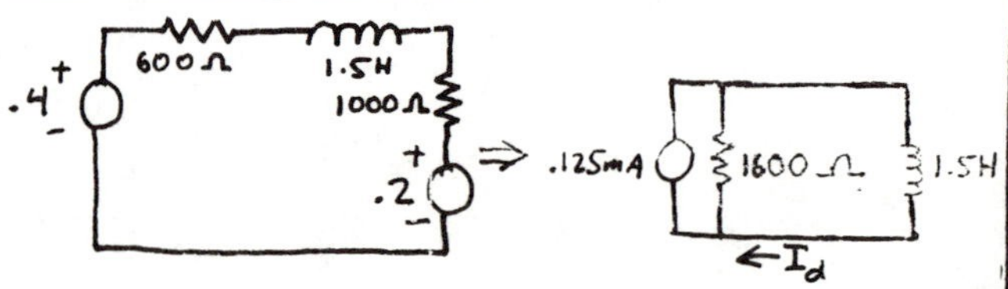

continues

3. Continued

As the final current (I_d) does not read .1mA, the circuit will a read a steady state condition of $I_d = .125$ mA, $V_d = .325$ V. Thus the circuit does not oscillate.

4. From the information given, the only known solution is that an ammeter was used in the current measurements. Even this is unsolvable unless it is assumed that the same meter (and scale) was used in all measurements:

$$Z_p = R_1 + a^2R_2 + R_m + j(X_1 + a^2X_2)$$

$$Z_s = Z_t = \frac{R_p}{a^2} + R_s + R_m + j\left(\frac{X_p}{a^2} + X_s\right)$$

or

$$a^2Z_s = R_p + a^2R_s + a^2R_m + j(X_p + a^2X_s)$$

then

$$a^2Z_s - Z_p = (a^2-1)R_m$$

$$R_m = \frac{a^2Z_s - Z_p}{a^2-1}$$

$$Z_p = (5.0 + j50.0)\ K\Omega$$

$$Z_s = 80 + j720$$

$$a = \frac{115 KV}{13.8 KV} \quad \text{then } R_m = 8.12\ \Omega$$

then

$$R_p + a^2R_s = 5000 - 8.12 \to 5000$$

$$X_p + a^2X_s = 50000$$

these generally split:

$$R_p = a^2R_s = \frac{5000}{2} = 2500\ \Omega$$

$$X_p = a^2X_s = \frac{50K}{2} = 25\ K\Omega$$

then

$$R_s = R_t = \frac{2500}{a^2} = 36\ \Omega$$

$$X_s = X_t = \frac{25K}{a^2} = 360\ \Omega$$

5.

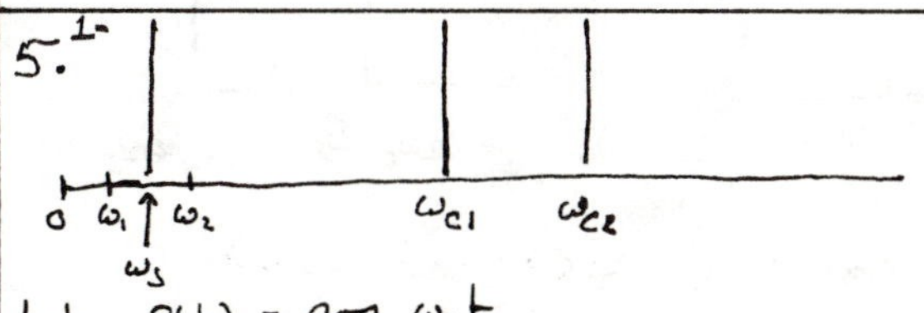

let $f(t) = \cos \omega_s t$

note: $\cos x \cos y = \frac{1}{2}\cos(x+y) + \frac{1}{2}\cos(x-y)$

$$f_1(t) = \frac{1}{2}\cos(\omega_{c1}+\omega_s)t + \frac{1}{2}\cos(\omega_{c1}-\omega_s)t$$

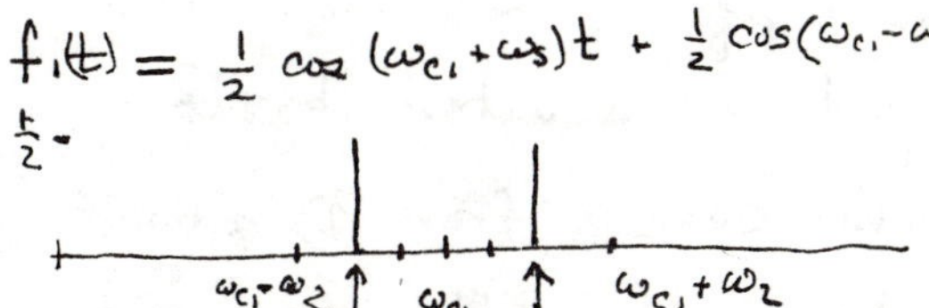

thus $f_1(t)$ contains the signal at half-magnitude shifted to two sidebands:

continues

5 continued

lower sideband from $\omega_{c_1} - \omega_2$ to $\omega_{c_1} - \omega_1$

upper sideband from $\omega_{c_1} + \omega_1$ to $\omega_{c_1} + \omega_2$

this is true for both (a) and (b)

$$f_2(t) = \frac{1}{4} \cos\left[(\omega_{c_1} + \omega_{c_2} + \omega_s)t + \phi\right]$$

$$+ \frac{1}{4} \cos\left[(\omega_{c_1} + \omega_{c_2} - \omega_s)t + \phi\right]$$

$$+ \frac{1}{4} \cos\left[(\omega_{c_1} - \omega_{c_2} + \omega_s)t - \phi\right]$$

$$+ \frac{1}{4} \cos\left[(\omega_{c_1} - \omega_{c_2} - \omega_s)t - \phi\right] \quad \text{①}$$

(a) $\omega_{c_1} = \omega_{c_2} = \omega_c$

$$f_2(t) = \frac{1}{4} \cos\left[(2\omega_c + \omega_s)t + \phi\right]$$

$$+ \frac{1}{4} \cos\left[(2\omega_c - \omega_s)t + \phi\right]$$

$$+ \frac{1}{4} \left[\cos(\omega_s t - \phi) + \cos -(\omega_s t + \phi)\right]$$

$$f_2(t) = \frac{1}{4} \cos\left[(2\omega_c t + \omega_s t + \phi)\right]$$

$$+ \frac{1}{4} \cos\left[(2\omega_c - \omega_s)t + \phi\right]$$

$$+ \frac{1}{2} \cos\phi \cos\omega_s t$$

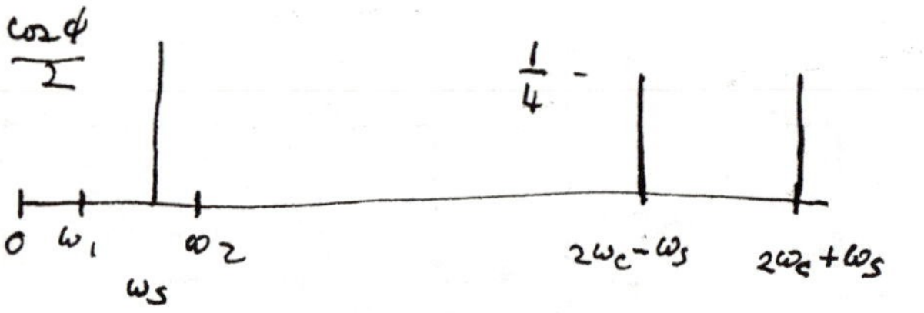

$f_2(t)$ contains $f(t)$ attenuated by a factor of $\frac{\cos\phi}{2}$, and the sidebands of f_1 shifted to $2\omega_c$ and attenuated by 2.

(b) In this case, from ① above we obtain the high frequency sidebands much as in (a), but at the center frequency of

$$\left[1 + |K|\right]\omega_{c_1}$$

There is also a low-frequency modulation at a frequency band at $\left[1 - |K|\right]\omega_{c_1} + \omega_1$ to

$$\left[1 - |K|\right]\omega_{c_2} + \omega_2$$

consider $f(t)$ to have a band spectrum from ω_1 to ω_2 as:

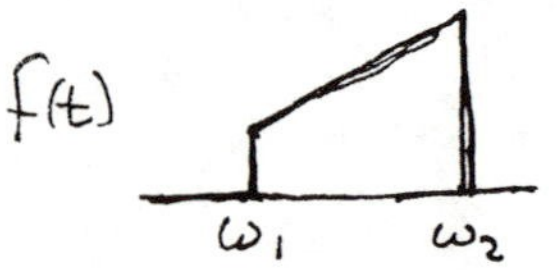

then depending on whether $\left[1 - |K|\right]\omega_{c_1}$ falls below ω_2, the lower side band will appear as the following cases:

1. $\omega_{c_1}\left|1 - |K|\right| > \omega_2$ let $\omega_{c_1}\left|1 - |K|\right| = \omega_0$

2. $\omega_1 < \omega_0 < \omega_2$

↑ this portion folds over

3. $\omega_0 < \omega_1$

↑ this portion folds over

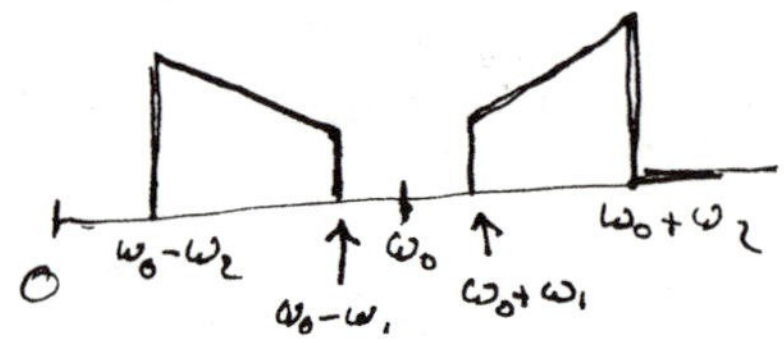

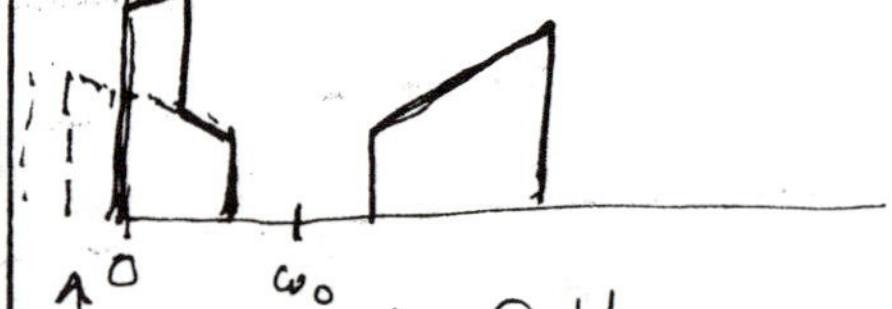

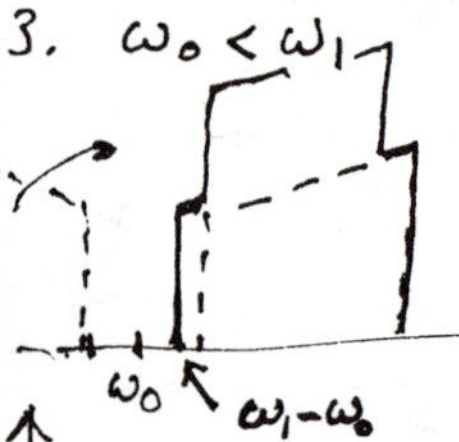

6. (a) $V_{RMS} = \sqrt{6^2 + \frac{100^2}{2}} = 70.96$ volts

(b) for a sinusoid of 100 volts peak the rms voltage is $70.7 = \frac{100}{\sqrt{2}}$

the average is $\frac{200}{\pi} = \frac{2}{\pi} \times 100$

for the meter to read RMS it is necessary for the meter to indicate $\frac{\pi}{2\sqrt{2}} V_{AVG}$

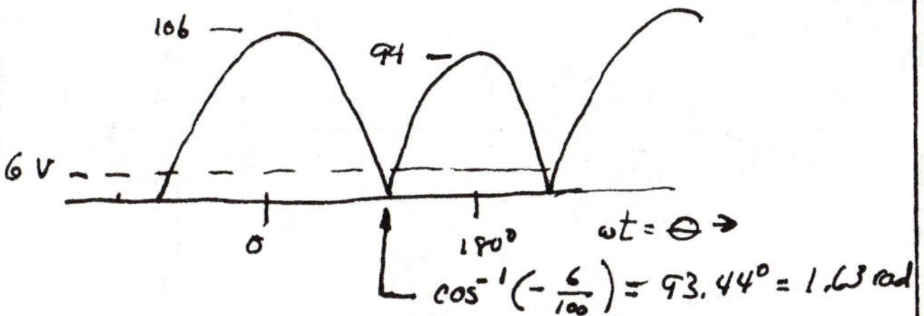

(c) $V_{avg} = \frac{1}{\pi}\int_0^{1.63}(6 + 100\cos\theta)d\theta - \frac{1}{\pi}\int_{1.63}^{\pi}(6 + 100\cos\theta)\,d\theta$

$= \frac{1}{\pi}\left[6 \times 1.63 + 100 \sin 1.63\right]$

$- \frac{1}{\pi}\left[6(\pi - 1.63) - 100 \sin 1.63\right]$

$V_{AVG} = 63.78$

$V_{reading} = \frac{\pi}{2\sqrt{2}} V_{AVG} = 70.83$

7.

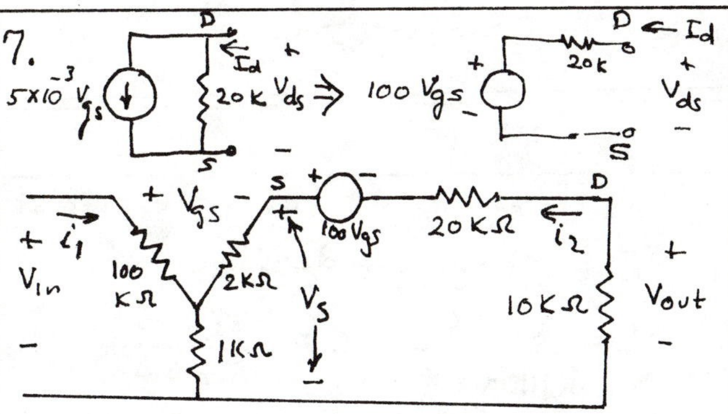

INPUT LOOP: $V_{IN} = 101K i_1 + 1K i_2$ ①

output Loop: $100 V_{gs} = 1K i_1 + 33K i_2$ ②

$V_S = 1K i_1 + 3K i_2$; $V_{gs} = V_{IN} - V_S = 100K i_1 - 2K i_2$ ③

③ into ② $10K i_1 = 233 i_2$ ④ ; $V_{out} = -10K i_2$ ⑤

①, ④ & ⑤ $A_V = \frac{V_{out}}{V_{IN}} = -2.98$ (a)

① & ④ $Z_{in} = \frac{V_{IN}}{i_1} = 144K$ (b)

8. first reduce the INNER LOOP

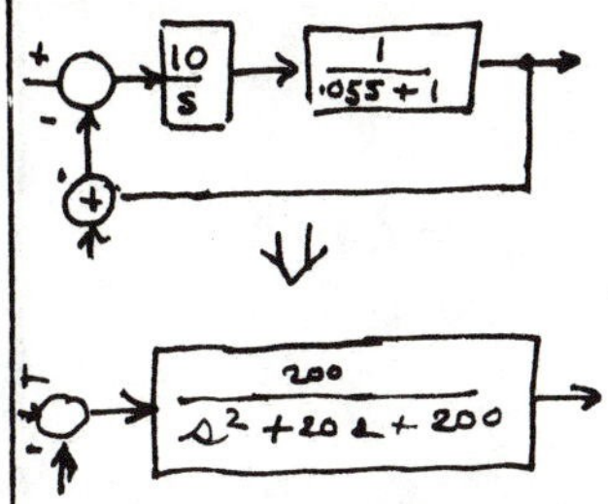

$\frac{\frac{10}{s}\frac{1}{.05s+1}}{1 + \frac{10}{s}\frac{1}{.05s+1}} = \frac{200}{s^2 + 20s + 100}$

$G = \frac{200K}{s^2 + 20s + 200}$ poles at $s = -10 \pm j10$

ROOT LOCI: $s^2 + 20s + 200 + 200K = 0$

$s = -10 \pm j10\sqrt{1 + 2K} = \sigma + j\omega$

$K = \frac{\omega^2 - 100}{200}$

ω	10	15	20	25	30
K	0	.625	1.5	2.625	4

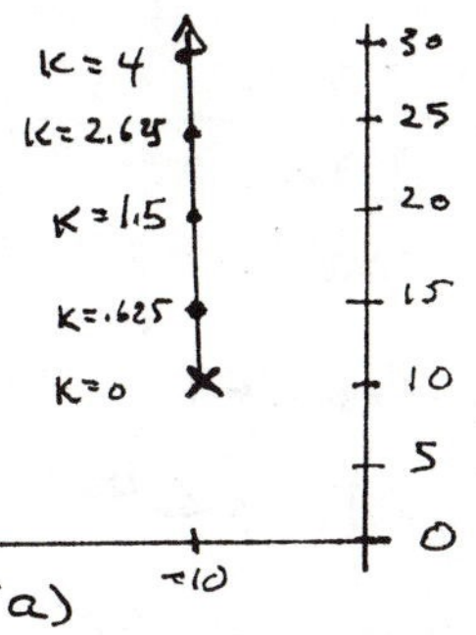

(a)

(b) the system is unconditionally stable

(c) $e_{SS} = \lim_{s \to 0} s \frac{R(s)}{1 + G(s)}$

$R(t) = u(t)$ $R(s) = \frac{1}{s}$

$\lim_{s \to 0} G(s) = K$

$e_{SS} = \lim_{s \to 0} s \frac{\frac{1}{s}}{1 + K} = \frac{1}{1 + K}$

the larger K, the smaller the steady-state error, but the lower the damping.

9. Arbitrarily define $V_1 = V_{ab}$, $V_2 = V_{bc}$ and $V_3 = V_{ca}$

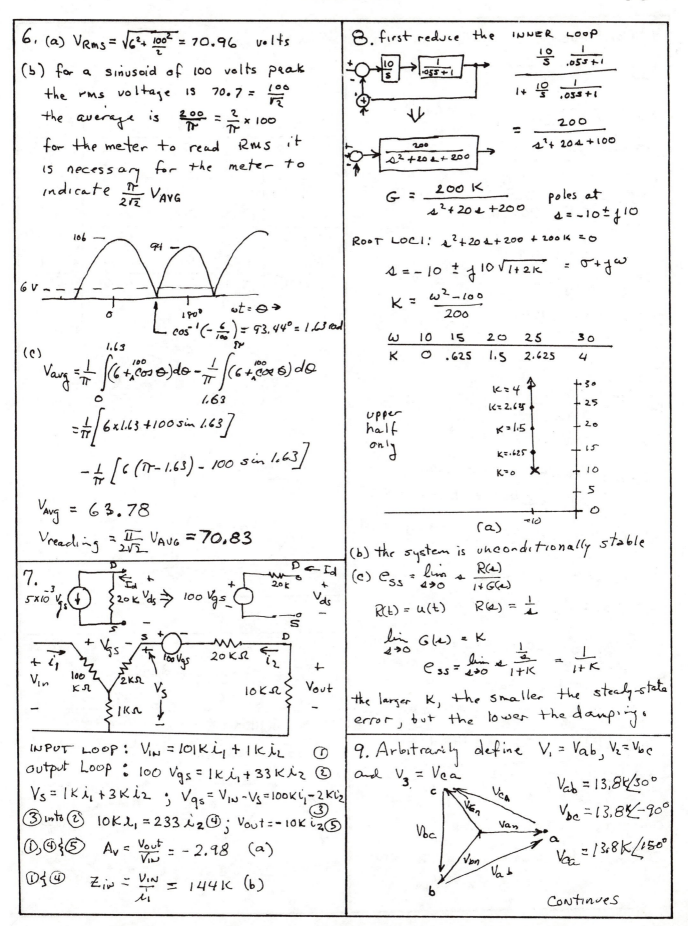

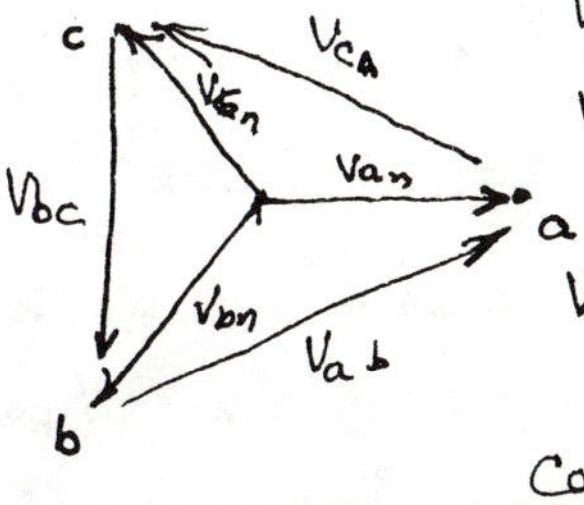

$V_{ab} = 13.8K\angle 30°$

$V_{bc} = 13.8K\angle -90°$

$V_{ca} = 13.8K\angle 150°$

Continues

9. Continued

$V_{an} = \frac{13.8KV}{\sqrt{3}} = 7.97K\angle 0^\circ$

$V_{bn} = 7.97K\angle -120^\circ$, $V_{cn} = 7.97K\angle +120^\circ$

unbalanced load

$\frac{V_{ab}}{Z_1} = I_{ab} = \frac{13.8K\angle 30^\circ}{(10+j10)K} = .943 - j.253$

$I_{bc} = \frac{13.8K\angle -90^\circ}{(5+j15)K} = -.828 - j.276$

$I_{ca} = \frac{13.8\angle 150^\circ}{j20} = .345 + j.598$

balanced load

$|I_{A\,bal}| = \frac{30/3}{(13.8/3) \times Pf} = 1.569$

$\angle I_A = -\cos^{-1} .8 = -38.7^\circ$

$I_{A\,bal} = 1.569\angle -38.7 = 1.255 - j.941$

$I_{B\,bal} = I_{A\,bal}\angle -120^\circ = -1.443 - j.616$

$I_{C\,bal} = I_{A\,bal}\angle 120^\circ = .188 + j1.558$

$I_a = I_{A\,bal} + I_{ab} - I_{ca}$

$= 1.853 - j1.770 = 2.563\angle -43.7^\circ A$

$S_a = V_{an} I_a^* = 20.42\angle 43.7^\circ = 14.77 + j14.11$ KVA

$I_b = I_{B\,bal} + I_{bc} - I_{ab}$

$= -3.214 - j.639 = 3.277\angle -168.8^\circ$ A

$S_b = V_{bn} I_b^* = 17.22 + j19.64$ KVA

$I_c = I_{C\,bal} + I_{ca} - I_{bc} = 2.787\angle 60.8^\circ$

$S_c = V_{cn} I_c^* = 11.36 + j19.09$ KVA

CURRENTS ARE: 2.56A, 3.28A & 2.79A (a)

Generator P = 14.77 + 17.22 + 11.36 KW

$\simeq$ 43.35 KW (b)

10. $K_A = \overline{B}$ $J_A = \overline{\overline{B} + C} = B \cdot \overline{C}$

$K_B = A$ $J_B = C$

$K_C = B$ $J_C = \overline{A + B} = \overline{A} \cdot \overline{B}$

A B C	K_A J_A	K_B J_B	K_C J_C	$A^+ B^+ C^+$
0 0 0	1 0	0 0	0 1	0 0 1
0 0 1	1 0	0 1	0 0	0 1 1
0 1 0	0 1	0 0	1 0	1 1 0
0 1 1	0 0	0 1	1 0	0 1 0
1 0 0	1 0	1 0	0 1	0 0 1
1 0 1	1 0	1 1	0 0	0 1 1
1 1 0	0 1	1 0	1 0	1 0 0
1 1 1	0 0	1 1	1 0	1 0 0

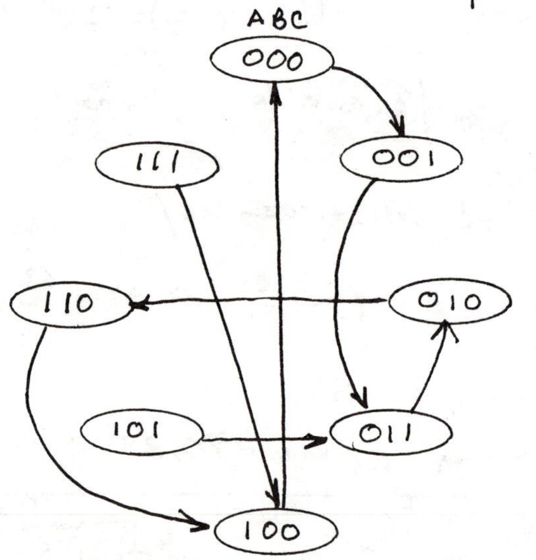

PART 2

11. the lower part of the circuit is a voltage divider;

$V_n = \frac{R\|10K}{R\|10K + 5K} 1.5 = \frac{3(1+\delta)}{5+3\delta}$ (1)

the ideal amplifier holds $V_I = V_n$

so

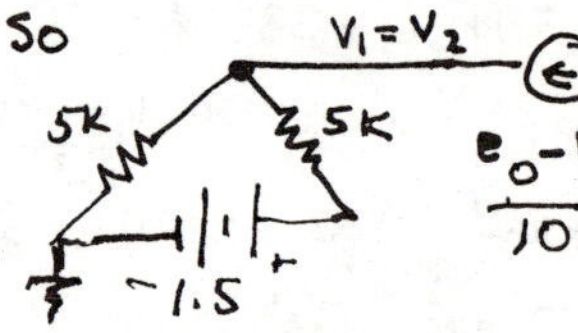

$\frac{e_o - V_n}{10K}$

sum CURRENTS At the top of the bridge

$\frac{e_o - V_n}{10K} + \frac{1.5 - V_n}{5K} = \frac{V_n}{5K}$; $e_o = -3 + \frac{15(1+\delta)}{5+3\delta}$ (2)

Continues

PART 2

11. Continued

THEN $e_o = \frac{6\delta}{3\delta + 5}$

δ	-1	-.8	-.6	-.4	-.2	0
e_o	-3	-1.8	-1.12	-.63	-.27	0

δ	0	.2	.4	.6	.8	1
e_o	0	.21	.38	.53	.65	.75

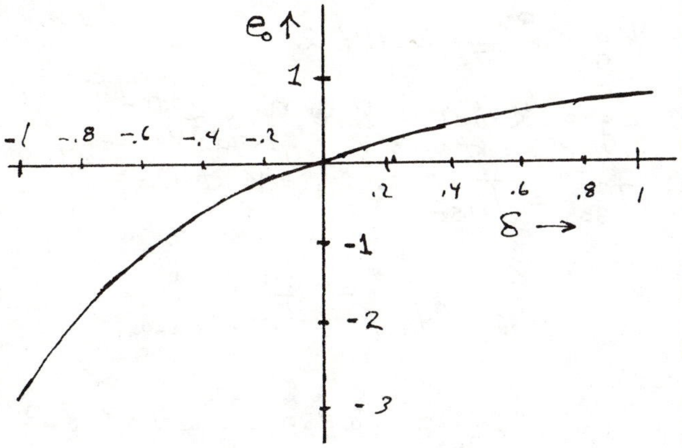

(a)

The resistance should have a linear change with the variable being measured, such as a resistance with a positive temperature coefficient of resistance to measure temperature:

$$R = R_o\left[1 + \alpha_o(T - T_o)\right]$$

$$= R_o(1 - \alpha_o T_o)\left[1 + \frac{\alpha_o T}{1 - \alpha_o T_o}\right]$$

$$\delta = \frac{\alpha_o}{1 - \alpha_o T_o}$$

for copper $\alpha_{20} = 3.9 \times 10^{-3}$; $\delta_{cu} = .0042$

Nichrome $\alpha_{20} = .017$; $\delta = .026$

the range of δ permitted depends on the accuracy required.

for an accuracy of n %, with $e_o|_{\delta=0} = 1.2\,\delta$ for the slope through the origin,

$$\left|\frac{1.25 - e_o}{1.2\delta}\right| < \frac{n}{100}$$

$$1.2 - \frac{n}{100} < \frac{6}{3\delta + 5} < 1.2 + \frac{n}{100}$$

for an accuracy of $n = 1\%$

$$\frac{-1}{101} \le .6\delta \le \frac{1}{99}$$

$$-.017 \le \delta \le .017$$

for an accuracy of $n = 5\%$

$$-\frac{5}{105} \le .6\delta \le \frac{5}{95}$$

$$-.079 \le \delta \le .088$$

12. Refer to Smith Chart p 6-22

reactances $z_A = \frac{-25}{50} = -.5$

$z_B = \frac{50}{50} = 1$

(A) the bottom of the Smith chart has $X = \infty$, which corresponds to an open line. Open lines with length $< .25\lambda$ will have negative reactance. Find $-.5$ on the periphery at .426 wavelengths and subract .25 wavelengths from the bottom for $L = .176\lambda$ with an open line.

(B) the top of the Smith Chart is $X = 0$, or a short. Positive reactances are obtained with shorted lines of $L < \lambda/4$. for $X = 1$, read on periphery $L = .125\lambda$ with a shorted line.

13. The following assumptions are made:

1. $V_A = 125$ in all cases
2. The power loss at 1150 RPM is equally divided among Armature resistance, field resistance and mechanical:

At 1150 RPM

$P_{in} - P_{out} = 125 \times 80 - 10 \times 746 = 2540$

$P_{mech\ loss} = \frac{2540}{3} = 847\ W$

$V^2/R_f = 847\ W \therefore R_f = 18.45\ \Omega$

$I_A^2 R_A = 846\ W \therefore R_A = .132\ \Omega$

R_A is constant throughout

3. The mechanical loss is proportional to speed [the alternative is that it is constant, but the speed change seems too large]

1300 RPM

$P_{mech\ loss} = \frac{1300}{1150} \times 847 = 957\ W$

800 RPM

$P_{mech\ loss} = \frac{800}{1150} \times 847 = 589\ W$

4. Speed control is by change of the field resistance, and the flux is proportional to field current. With V_A constant $\phi \propto \frac{1}{R_f}$

(a) $n = 1300$ RPM

$P_{out} = \frac{1300}{1150} \times 7460 = 8344\ W$

$P_{Airgap} = P_{out} + P_{mech}$
$= 8344 + 957 = 9390$

$P_{Airgap} = E_g I_A \therefore E_g = \frac{9390}{I_A}$

$V_A = I_A R_A + E_g = .132 I_A + \frac{9390}{I_A} = 125$

solve for I_A:

$I_A = \frac{125 \pm \sqrt{125^2 - 4 \times .132 \times 9390}}{2 \times .132}$

$= 865,\ \boxed{82.2}$

$E_g = \frac{9390}{I_A} = 114.2$

$E_g = K\phi\Omega = K'\phi n : K'\phi = .0878$

$K'\phi = K'' V_A : K'' = .7 \times 10^{-3} V_A$

$T_{1300} = T_{1150} : T = K_T \phi I_A = \frac{K_T'}{R_f} I_A$

$\frac{K_T'}{R_f} I_A \Big|_{1150} = \frac{K_T'}{R_f} I_A \Big|_{1300}$

$R_{f\,1300} = 18.45 \times \frac{82.2}{80} = 18.96\ \Omega$

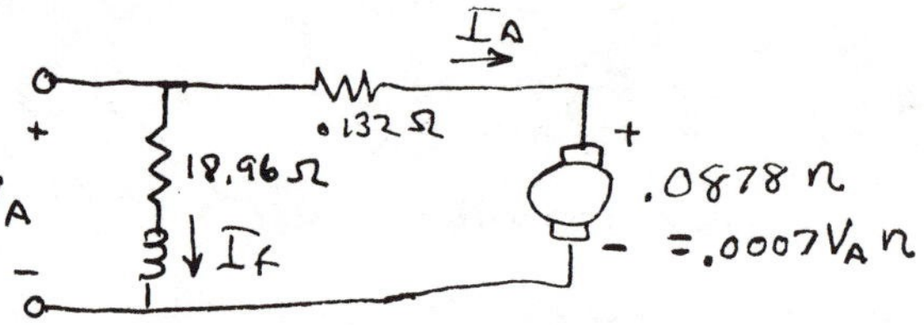

(b) $n = 800$ RPM [following part (a)]

$P_{AIRGAP} = \frac{800}{1150} \times 7460 + 589 = 5779$

$I_A = \frac{125 \pm \sqrt{125^2 - 4 \times .132 \times 5779}}{2 \times .132}$

$= 898,\ \boxed{48.75}$

$E_g = \frac{5779}{48.75} = 118.5 = K'\phi \times 800$

$E_g = .148\ n$
$= .0012\ V_A n$

$R_f = 18.45 \times \frac{48.75}{80} = 11.24\ \Omega$

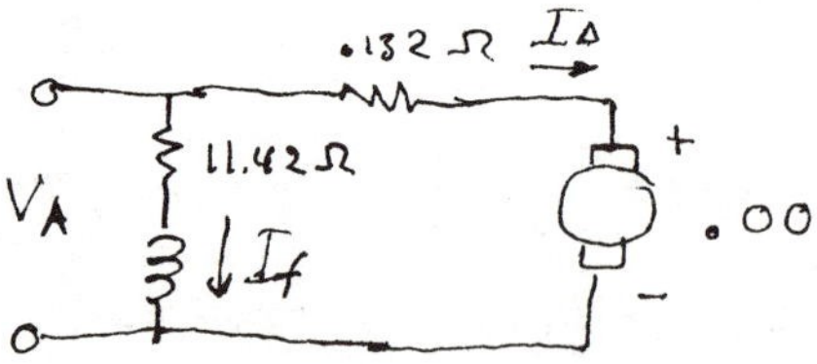

14. For the timing circuit:

$\omega RC = 377 \times 10^{5} \times 10^{-7} = 3.77$

$V_C(\text{s.s.}) = \frac{V_s}{1+j3.77} = \frac{220\sqrt{2}\angle 0}{1+j3.77}$

$V_C(t)_{\text{s.s.}} = 79.77 \cos(377t - 75.1°)$ ↖ 1.312 rad.

diac begins each half cycle at zero volts, so $v_C(0) = 0$

$v_C(t) = v_C(t)_{ss} + A e^{-t/RC}$

$0 = 79.77 \cos(-75.1°) + A$

$A = 20.45$

$v_C(t) = 20.45 e^{-100t} + 79.77 \cos(377t - 75.1°)$

at t_f, $v_C(t_f) = 4$ volts

Iteration formula

$t_{f2} = \frac{1}{377}\left[1.312 + \sin^{-1} \frac{4 - 20.45 e^{-100 t_{f1}}}{79.77}\right]$

t_f iterations: milliseconds

0, 2.93, 3.14, 3.15, 3.15, 3.15 ....

$T_{\text{half-cycle}} = \frac{1}{120} = 8.33 \times 10^{-3}\,\text{s} = 8.33\,\text{ms}$

the firing angle is $\frac{3.15}{8.33} \times 180° = 68°$

Conduction angle is $180° - 68° = 112°$

15. $V_s = 5 I_d + V_d$

V_d	−1.5	−1	−.5	0	.5	1.0	1.5
I_d	.079	.033	.0054	0	.0054	.033	.079
V_s	−1.11	−.835	−.473	0	.527	1.165	1.895

interpolate to find Vd at $V_s = -1$

$\frac{V_d + 1}{-1.5 + 1} = \frac{-1 + .835}{-1.11 + .835}$

$V_d \approx -1.3$

interpolate to find Vd at $V_s = 1$

$\frac{V_d - .5}{1 - .5} = \frac{1 - .527}{1.165 - .527}$

$V_d = .871$

V_s	−1	−.835	−.437	.527	+1
t	−90°	−56.6°	−25.9°	31.8°	90°
V_d	−1.3	−1	−.5	.5	.871

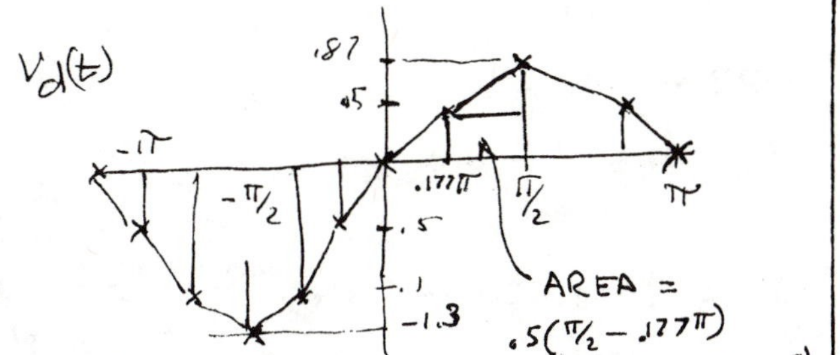

$.5(\pi/2 - .177\pi) + \frac{1}{2}(\pi/2 - .177\pi)(.871 - .5)$

$= \frac{1}{2}(\pi/2 - .177\pi)(.871 + .5)$

This has an average value calculated by trapezoidal areas

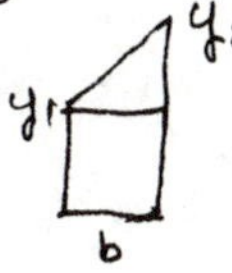

$A \doteq b y_1 + \frac{1}{2} b (y_2 - y_1)$

$= \frac{1}{2} b (y_1 + y_2)$

the area is calculated from doubling the area from $-\pi/2$ to $+\pi/2$

the average found is −.111 volts

$V_d(t) = V_{DC} + \sum a_n \cos n\omega t + \sum b_n \cos n\omega t$

eq (1):

$V_d(t) + .111 = a_1 \cos t + a_2 \cos 2t + a_3 \cos 3t + b_1 \sin t + b_2 \sin 2t + b_3 \sin 3t$

6 coefficients can be calculated from the 5 points in the above table + the value at zero

t (degs)	−90	−56.6	−25.9	0	31.8	90
V_d+.111	−1.189	−.889	−.389	.111	.611	.982
cos t	0	.55	.90	1	.85	0
cos 2t	−1	−.39	.62	1	.44	−1
cos 3t	0	−.98	.21	1	−.09	0
sin t	−1	−.83	−.44	0	.53	1
sin 2t	0	−.92	−.79	0	.9	0
sin 3t	1	−.18	−.98	0	1	−1

15 continued.

Construct the 6 equations from eq. 1 for each column of the last table. From left to right:

$-1.189 = -a_2 - b_1 + b_3$ ②

$-.889 = .55a_1 - .39a_2 - .98a_3 - .83b_1 - .92b_2 - .18b_3$ ③

$-.389 = .9a_1 + .62a_2 + .21a_3 - .44b_1 - .79b_2 - .98b_3$ ④

$.111 = a_1 + a_2 + a_3$ ⑤

$.611 = .85a_1 + .44a_2 - .09a_3 + .53b_1 + .96b_2 + b_3$ ⑥

from ② and ⑥ determine that

$a_2 = .104$ ⑦

$b_3 = b_1 - 1.086$ ⑧

from ⑤ and ⑦

$a_3 = .007 - a_1$ ⑨

using ⑦, ⑧ and ⑨ in ③:

$-.678 = a_1 - .66b_1 - .601b_2$ ⑩

using ⑦, ⑧ and ⑨ in ④:

$-2.232 = a_1 - 2.058b_1 - 1.145b_2$ ⑪

using ⑦, ⑧ and ⑨ in ⑥

$1.757 = a_1 + 1.628b_1 + .957b_2$ ⑫

from ⑩ & ⑪

$2.857 = 2.57b_1 + b_2$ ⑬

from ⑪ & ⑫

$1.898 = 1.754b_1 + b_2$ ⑭

$b_1 = 1.175, \quad b_2 = -.163$

from ⑩, ⑪ or ⑫ $a_1 = 0$

from ⑨ $a_3 = .007$

from ⑧ $b_3 = .089$

then

$V_d \approx .111 + 1.175 \sin t$

$+ [.104 \cos 2t - .163 \sin 2t]$

$+ [.007 \cos 3t + .089 \sin 3t]$

d.c. $V_{dc} = .111$

$f = \frac{1}{2\pi}$ $V_{\frac{1}{2\pi}} = \frac{1.175}{\sqrt{2}} = .831$ 100%

2d harm $V_{\left(\frac{2}{2\pi}\right)} = \sqrt{\frac{.104^2}{2} + \frac{.163^2}{2}} = .137$ 16%

3d harm $V_{\left(\frac{3}{2\pi}\right)} = \sqrt{\frac{.007^2}{2} + \frac{.089^2}{2}} = .063$ 7.6%

16. $K = \overline{\overline{Q_1 \cdot \overline{Q_2}} \cdot Q_3} = \overline{Q_3} + Q_1 \cdot \overline{Q_2}$

$J = Q_2$

$D = \overline{\overline{Q_1} + Q_2} = Q_1 \cdot \overline{Q_2} = \overline{Q_2} Q_1$

$R = \overline{\overline{Q_2} + \overline{Q_3}} = Q_3 \cdot Q_2$

$S = \overline{Q_2}$

Q_3	Q_2	Q_1	K	J	D	S	R	Q_3^+	Q_2^+	Q_1^+
0	0	0	1	0	0	1	0	0	0	1
0	0	1	1	0	1	1	0	0	1	1
0	1	0	1	1	0	0	0	1	0	0
0	1	1	1	1	0	0	0	1	0	1
1	0	0	0	0	0	1	0	1	0	1
1	0	1	0	0	1	1	0	1	1	1
1	1	0	0	1	0	0	1	1	0	0
1	1	1	1	1	0	0	1	0	0	0

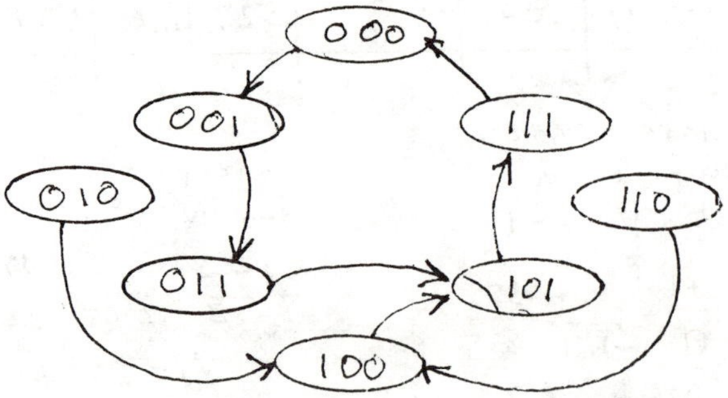

17 $900I_E + V_{CE} + 1000I_C = 24$

set $V_{CE} = \frac{24}{3} = 8\text{ V}$

$$I_C = I_{CO} + \frac{\beta}{1+\beta} I_E$$

$I_{CO} \approx .1\text{ mA}$ $\quad I_C \sim \frac{24-8}{2000} = 8\text{ mA}$

ignore I_{CO}

$$I_C = \frac{35}{36} I_E$$

$$\left(900 + \frac{35}{36}1000\right) I_E = 24 - 8$$

$$I_E = 8.55\text{ mA}$$

$$V_E = 8.55\text{mA} * .9\text{K}\Omega = 7.69\text{ V}$$

$$V_B = 7.69 + .2 = 7.89$$

V_E ↑ (7.69); ↑ V_{be} (germanium) (.2)

$$I_b = \frac{I_e}{1+\beta} = .238\text{mA}$$

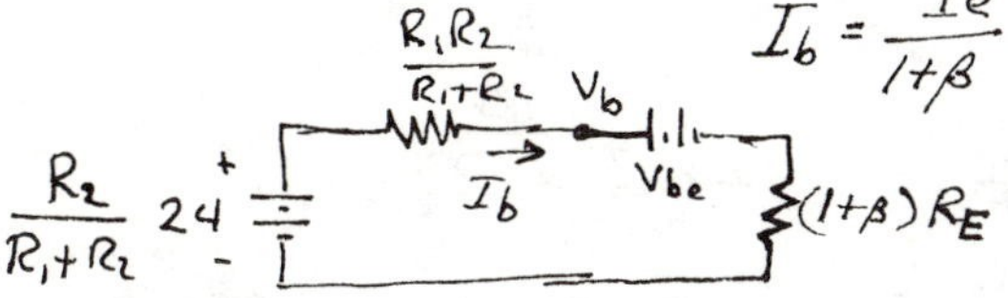

$$\frac{R_2}{R_1+R_2} 24 = .238 \times 10^{-3} \frac{R_1R_2}{R_1+R_2} + 7.89 \quad (1)$$

bias stability: $(1+\beta)R_E \gg \frac{R_1R_2}{R_1+R_2}$

note: see eq. 8.77, p 8-19

$$(1+\beta)R_E = 32.4\text{ K}$$

as a compromise, set $\frac{R_1R_2}{R_1+R_2} = 3.9\text{K}$

then $\frac{R_2}{R_1+R_2} 24 = .238 \times 3.9 + 7.89 = 8.82$

$$\frac{R_2}{R_1+R_2} = \frac{8.82}{24} = .367$$

then $.367R_1 = 3.9\text{ K}$ $\quad R_1 = 10.35\text{K}$

choose a practical $R_1 = 10\text{K}$

then in eq (1)

$$\frac{R_2}{R_1+R_2} = .365 \quad \left\{ \quad R_2 = \frac{.365}{1-.365} R_1 = 5.7\text{K} \right.$$

small-signal mid-frequency circuit

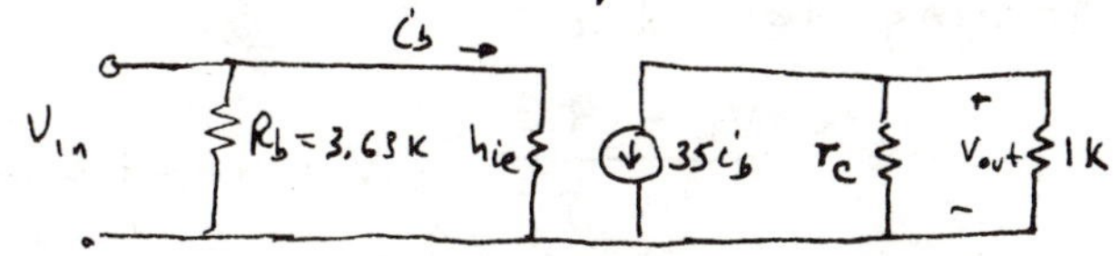

input impedance: $h_{ie} \| R_b = \frac{R_1R_2}{R_1+R_2}$

typically: $100 < h_{ie} < 700\,\Omega$

then: $97 < Z_{in} < 580\,\Omega$

output impedance: $1\text{k} \| r_c$

typically: $5\text{K}\Omega < r_c < 20\text{K}\Omega$

then $833\Omega < Z_o < 952\,\Omega$

Voltage Gain:

$$A_v = -\frac{\beta}{h_{ie}} \times (r_c \| R_L)$$

$$= -\frac{35}{h_{ie}} Z_o$$

then

$$\frac{35\,Z_{o\,MIN}}{h_{ie\,MAX}} = 42 < |A_v| < 333 = \frac{35\,Z_{o\,MAX}}{h_{ie\,MIN}}$$

18.

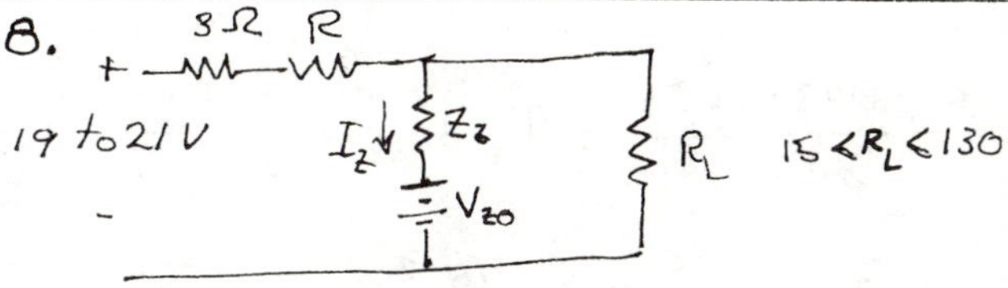

Try 16V, 50 W zener diode p 9-29

$$V\Big|_{1/3\text{ rated current}} = V_{ZO} + Z_Z \frac{\text{rated current}}{3}$$

$V = 16\text{V}$, $Z_Z = 1.6\,\Omega$, $I_{ZK} = 5\text{ mA}$

rated current $= \frac{50\text{W}}{16\text{V}}$

then $V_{ZO} = 16 - \frac{50\text{W}}{3 \times 16\text{V}} \times 1.6\,\Omega = 14.3\text{ V}$

At lowest supply voltage and highest load, $I_Z > I_{ZK}$

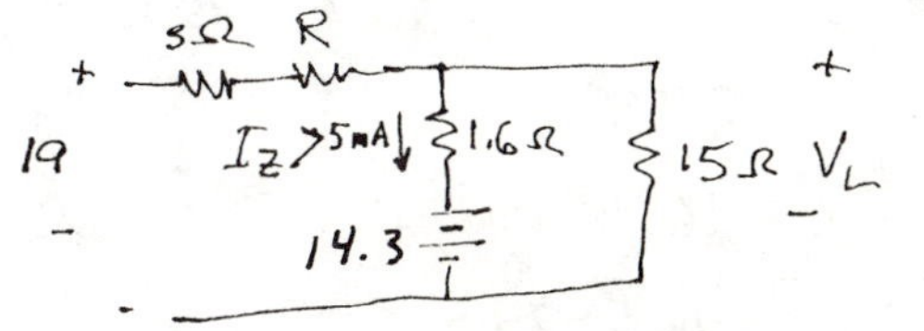

18 continued

solving for I_z:

$$I_z = \frac{V_L - 14.3}{1.6} \qquad V_L = 14.3 + 1.6 I_z$$

$$\frac{19 - V_L}{3+R} = I_z + \frac{V_L}{15} \quad \text{then}$$

$$\frac{4.7}{3+R} - \frac{14.3}{15} = I_z\left[\frac{16.6}{15} + \frac{1.6}{3+R}\right]$$

as $I_z > .005$

$$\frac{4.7}{3+R} - \frac{14.3}{15} > .005\left[\frac{16.6}{15} + \frac{1.6}{3+R}\right]$$

then $R < 1.893\,\Omega$

At maximum supply voltage and minimum load, $I_z < I_{ZMAX} = \frac{50W}{16V}$

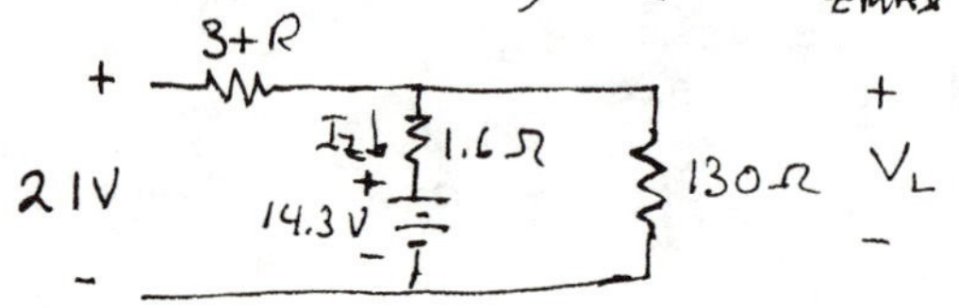

$$\frac{21 - (14.3 + 1.6 I_z)}{3+R} = I_z + \frac{14.3 + 1.6 I_z}{130}$$

$$I_z = \frac{\frac{6.7}{3+R} - \frac{14.3}{130}}{\frac{131.6}{130} + \frac{1.6}{3+R}} < \frac{50}{16} = 3.125$$

at $R = 0$ $\qquad I_Z = 1.374$ ok

use $R = 0$

them at min $V_s = 19$, max load $\rightarrow 15\,\Omega$

$$\frac{4.7}{3} - \frac{14.3}{15} = I_z\left[\frac{16.6}{15} + \frac{1.6}{3}\right]: I_z = .374$$

$$V_L = 14.9\text{ V}$$

at max $V_s = 20$, min load $\rightarrow 15\,\Omega$

$I_z = 1.374$

$V_L = 16.5$ volts

Regulation: $\frac{16.5 - 14.9}{14.9} \times 100 = 10.7\%$

with $R = 1.5\,\Omega$

$V_{L\,min} = 14.4$

$V_{L\,max} = 15.9$

Regulation $\rightarrow 10.4\%$

while this doesn't look too good, the regulation is 30% without the zener diode.

Alternate solution: Design to specify minimum zener ratings.

Ignore R_Z, $I_{ZK} \rightarrow 0$

$$\frac{V_{min} - 15}{R+3} = I_z + \frac{15}{15} \qquad I_z \rightarrow 0$$

$$R + 3 = 19 - 15 - 1 = 3 \qquad R = 0$$

$$\frac{V_{max} - 15}{3} = I_{Zmax} + \frac{15}{130} \qquad I_{Zmax} = 1.885$$

Zener RATINGS: $15\text{V} \times 1.885\text{ A}$

28.3 W

$R = 0$

This ignores the regulation problem

19. Use line base

line: $\frac{115\text{ KV}}{\sqrt{3}} \times 240\text{ A} \times 3\text{ phases} = 47.8\text{ MVA}$

transformer: Z_{tr} (47.8 MVA base)

$$= (.01 + j.08)\frac{47.8}{50} = .01 + j.076$$

$$Z_{total} = .11 + j.176 = .176\angle 86.7°$$

$$I = \frac{1}{.176}\angle -86.7 = 5.67\angle -86.7°\text{ p.u.}$$

$$I = 240 \times 5.76\angle -86.7°$$

$$\simeq 1361\angle -86.A \text{ A/phase}$$

20.

$$s^3X = R - 75s^2X - 1000sX - 6000X$$

$$R = (s^3 + 75s^2 + 1000s + 6000\ X$$

$$C = 20sX + 6000X = 20(s+30)X$$

$$\frac{C}{R} = \frac{20\ (s+30)}{s^3 + 75s^2 + 1000s + 6000}$$

measurement: $s^2 + 2\zeta\omega_n s + \omega_n^2$

$$= s^2 + 2 \times .75 \times 10\, s + 100$$

$$\begin{array}{r} s + 60 \\ s^2 + 15s + 100 \overline{)\, s^3 + 75s^2 + 1000s + 6000} \\ \underline{s^3 + 15s^2 + 100s} \\ 60s^2 + 900s + 6000 \\ \underline{60s^2 + 900s + 6000} \\ 0 \end{array}$$

(a)

$$\frac{C}{R} = \frac{20(s+30)}{(s+60)(s^2 + 15s + 100)}$$

Assume unity feedback

$$G = \frac{C/R}{1 - C/R} = \frac{20(s+30)}{s^3 + 75s^2 + 1000s + 6000 - 20s - 600}$$

$$= \frac{20(s+30)}{s(s^2 + 75s + 980)}$$

(b)

$$G = \frac{20\ (s+30)}{s\ (s+58.15)(s+16.85)}$$

$$\frac{R}{X} = s^3 + K_1 s^2 + K_2 s + K_3$$

$$\begin{array}{r} s^2 + (K_1 - 60)s + [K_2 + 3600 - 60K_1] \\ (s+60) \overline{)\, s^3 + K_1 s^2 + K_2 s + K_3} \\ \underline{s^3 + 60s^2} \\ (K_1 - 60)s^2 + K_2 s \\ \underline{(K_1 - 60)s^3 + 60(K_1 - 60)s} \\ K_2 + 3600 - 60K_1\, s + K_3 \end{array}$$

then $\omega_n^2 = 100 = K_2 + 3600 - 60K_1$

and $K_3 = 60 \times 100 = 6000$ unchanged

$$20\zeta = (K_1 - 60)$$

$$K_1 = 60 + 20\zeta$$

$$K_2 = 60K_1 - 3500$$

$$= 3600 + 1200\zeta - 3500$$

$$= 100 + 1200\zeta$$

$\zeta > 0$ $\quad K_2 > 100$

$K_1 > 60$